Stripe Press
Ideas for progress
South San Francisco, California
press.stripe.com

The Scaling Era
An Oral History of AI,
2019–2025

Dwarkesh Patel
with Gavin Leech

Published in Belgium
by Stripe Press / Stripe Matter Inc.

Stripe Press
Ideas for progress
South San Francisco, California
press.stripe.com

ISBN 978-1-953953-55-1 (print)
ISBN 978-1-953953-56-8 (ebook)

Also available in audiobook.

Library of Congress Control Number:
2024952033

"The Bitter Lesson" by Richard Sutton,
"The Scaling Hypothesis" (excerpt) by Gwern
Branwen, and "On the Broad Success of
Transformers" by Nostalgebraist reprinted
with kind permission by the authors.

"T-Rex as: 'The Computer Scientist,' Part 2"
by Ryan North reprinted with kind permission by the artist.

This book contains footnotes and endnotes.
Footnotes are denoted with a silver bullet.
Endnotes are denoted with a number.

Preface

There's a Sherlock Holmes story that captures our relationship with large language models.

A new client comes to Baker Street. With a single glance, Holmes rattles off the man's life story: that he lived in China, that he is a Freemason, that he writes a lot. The client, astonished, asks how Holmes knows all this. In great detail, Holmes explains the series of deductions that led him to his conclusions. The client responds, "I thought at first that you had done something clever, but I see that there was nothing in it, after all."[1]

Because we witness a series of small improvements to LLMs' intelligence piecemeal over years, and because we *think* we understand how they work, it's very easy to have the same reaction: "There's nothing in them, after all." That dismissiveness highlights why it's important to step back and see how remarkable their progress has been, and how little we understand them.• Let's try to see the last six years afresh.

A new technology arrives—call it *the thing*. Broadly speaking, we made it by having it read the entire internet until it learned how to respond when we talk to it. Through some 15 trillion rounds of trial and error, it wound up pretty smart.•• We don't really know how the resulting model works. We didn't design it so much as grow it.

MODEL
The AI system produced by training an architecture on data; a program that has learned to perform specific tasks.

Unlike all past software, the thing can speak with flawless grammar on any topic in any major world language. It can write in any literary style. It can "see" and discuss what it sees. It "knows" facts about millions of people.••• If you ask it to give smarter answers, or otherwise imply that you yourself are smart, it replies more thoughtfully.[2]

• In this book, I speculate both about how current systems work and how future systems will perform. But speculation isn't always self-indulgent or irrational. As Alan Turing noted in his 1950 paper "Computing Machinery and Intelligence," "the popular view that scientists proceed inexorably from well-established fact to well-established fact, never being influenced by any unproved conjecture, is quite mistaken. Provided it is made clear which are proved facts and which are conjectures, no harm can result."

•• This refers to the pretraining token total for Llama 3.1 405B. See Meta, "Introducing Meta Llama 3."

••• In the past, careful writers placed cognitive verbs like "read" and "understand" in quotes, since all that was happening inside the system was data processing—that is, syntactic operations without human semantics—and they needed to caution people not to expect too much. Now, as LLMs reach human-level performance in some domains, the need for quotation marks is less clear. I'll omit them in the rest of this book.

Developing it solved several open problems in AI, from learning complex tasks without explicit human labeling to processing data from multiple senses, such as video and audio, simultaneously.• It only acts when prompted, and the only action it ever takes is to predict what should come next.••

The thing is already plainly superhuman on a few dimensions: reading and writing speed, breadth of knowledge, number of languages spoken fluently, and accuracy at predicting the next word in a sentence.••• Despite not being intentionally trained for it, it's an incredibly effective general compression algorithm.[3] In theory, with the right data, it could perform any task.••••

It is also blatantly subhuman in other domains: the sheer number of examples it needs to permanently learn new skills, its failure at precise symbol manipulation tasks, and its lack of certain kinds of common sense, episodic memory, and the ability to seek out new evidence and be corrected.[4] People who work with it know to constantly

LEARNING
The process of adjusting weights in a model after it processes data, enabling improved predictions based on past performance.

PROMPT
The input provided by the user, typically a query or instruction that the model responds to.

TRAINING
The process of updating a model on data to improve its performance through trial and error. For LLMs, training begins with pretraining, where the model is exposed to a dataset, predicts the next part of the training data (autoregression), and is automatically updated based on the accuracy of its predictions. This builds the model's general language understanding and encodes a wide range of facts and relationships. Post-training encompasses a variety of techniques including instruction tuning, supervised fine-tuning, reinforcement learning from human feedback (RLHF), and direct preference optimization, which refine the model's ability to engage in extended dialogue, assist users, avoid malicious or off-policy outputs, and acquire task-specific knowledge.

• In 2006, the AI researcher Tom Mitchell listed nine open problems in the field, including "Can unlabeled data be helpful for supervised learning?" and "How can we transfer what is learned for one task to improve learning in other related tasks?" Four of these problems, including those related to unlabeled data and task transfer, have since been completely solved. Mitchell, "Discipline of Machine Learning"; Leech, "Mitchell's Open Problems."

•• However, these systems can do more than predict words. Since 2021's WebGPT, they've been able to take other actions, such as using tools and searching the web. The remarkable SayCan system uses an LLM to produce plans for a robot to execute. Nakano et al., "WebGPT"; Ahn et al., "Do As I Can, Not As I Say."

••• I encourage you to confirm these by inspection; I think it's that obvious. For a closer look at LLMs' superior ability to complete sentences, see Shlegeris et al., "Next-Token Prediction."

•••• A 2019 paper proves that Transformers are computationally universal. However, the result unrealistically assumes hard attention, which picks a single part of the input to pass on, whereas all actual LLMs use soft attention, which involves averaging over all inputs. Additionally, computability tells us little about the practicality of learning or running a particular program. Other theoretical results find that fixed-depth Transformers have hard limits, particularly for large inputs. But LLM systems now involve much more than just a pretrained Transformer, and some of these post-training enhancements and tool augmentations are known to expand the computability class. So we end up in a complex situation. Pérez et al., "Turing Completeness"; Dziri et al., "Faith and Fate"; Yehudai et al., "When Can Transformers Count to N?"; Merrill and Sabharwal, "Expressive Power of Transformers."

doubt its output. It's like having a brilliant but somewhat deluded and amnesic coworker.

Throughout the 20th century, we got used to computers being limited and uneven: simultaneously superhumanly fast and precise, with perfect recall, but also completely unable to understand natural language, infer anything from nonsymbolic data like images, or apply common sense to even slightly ambiguous requests. But we haven't gotten used to the thing's unevenness, so we tend to round it down (to a "stochastic parrot") or up (to something akin to a person, a replacement for expertise).[5]

Unlike past systems, the thing is massively multitask: one large model we can use for most functions that outperforms most single-task systems at their own game.• This means that, for example, a car dealership AI is perfectly happy to answer your questions about advanced mathematics.[6]

Multiple US corporations created their own versions of the thing using a large fraction of the world's copyrighted data.[7] One of them open-sourced its version after spending hundreds of millions of dollars making it.•• As a result of other companies developing the thing, a producer of video game hardware briefly became the world's most valuable business.[8]

It gets called a "large" model, yet its weights fit on a thumb drive.•••

Even the most powerful versions are freely available for casual use. It takes about 10 seconds to have it read and discuss a long book with you. In short bursts, it can

WEIGHT
A parameter that defines the strength of a connection between two units in a neural network; where the algorithms performed on inputs to produce outputs are defined. Metaphorically, weights are like the synapses in the brain.

• See, for example, Goyal et al., "Learning Activation Functions." There are exceptions, notably the deep RL lineage that has made progress on hard scientific tasks.

•• One unconfirmed estimate puts the total hardware used for training Meta's Llama 3.1 405B at 37 million H100-hours (at least 97 days on 16,000 H100 GPUs). At retail rates for these GPUs, this would cost around $75 million. But this omits many factors, including wholesale discounts and salaries for more than 200 research staff. Plappert et al., "Thoughts on Llama 3."

••• The full weights of Meta's Llama 3.1 405B model, for example, are only 820 GB.

code as well as many human professionals. Google is now using it to generate around 25 percent of its new code and merging it into its codebase unedited.•

Even so, most people don't seem that interested in the thing. Currently, only 5 percent of companies use it (officially).•• The market doesn't seem to expect it to become superhuman.[9] The leading company building it was on track to lose $5 billion in 2024.[10]

Some say it is just a compressed version of the internet—although it occasionally generates information that isn't on the internet.••• Sometimes it restates material out of context, like when it advised someone to eat rocks for their nutritional benefits.[11] We always assumed robots would be like computers: rigid, logical, and unable to create.•••• Instead, we find it hard to *stop* the thing from making stuff up.•••••

• In a 2022 blog post, Google engineers noted that 3 percent of new code at Google was generated by their code model and merged by a human without modification. An update from 2024 cites a complicated metric suggesting 50 percent. Google CEO Sundar Pichai claimed 25 percent in a 2024 earnings call. Tabachnyk and Nikolov, "ML-Enhanced Code Completion"; Chandra and Tabachnyk, "AI in Software Engineering at Google"; Pichai, "Q3 Earnings Call."

•• Some large companies have actually tried to ban the internal use of external AIs for fear of data leaks. *The Economist*, "What Happened to the Artificial Intelligence Revolution?"

••• One way in which an LLM is straightforwardly *not* just compressing the internet is its post-training, which comes after its being pretrained on the internet corpus. Most notably, most LLMs undergo millions of rounds of reinforcement learning from human feedback (RLHF). Figuratively, RLHF is like giving the model electroshocks to punish undesirable outputs and treats to reward desirable outputs. This process makes an LLM much less likely to go off topic or say unnerving things, but is also known to damage LLMs' calibration, or the accuracy of its confidence in its own statements. Chiang, "ChatGPT Is a Blurry JPEG of the Web"; Leopold Aschenbrenner (@leopoldasch), "Interesting that RLHF seems to worsen GPT-4's calibration," X, March 14, 2023, https://x.com/leopoldasch/status/1635713431088275457.

•••• We also assumed they'd be robots with physical bodies.

••••• See, for instance, hallucination, a vague term for when an AI generates plausible false information, often coupled with the model insisting that it is true. A more precise term is confabulation, defined as LLMs "fluently making claims that are both wrong and... sensitive to irrelevant details." Some researchers claim that hallucination is inherent to LLMs and is unfixable without the model interacting with the world. Farquhar et al., "Detecting Hallucinations."

Some of its creators talk about it in metaphysical terms: "We're creating God."[12] Or in uncanny terms: "They just want to learn."• Some activists, anticipating disaster, have called for a ban on systems more powerful than the current version.[13]

The thing can be made to subvert its intensive ethics training just by talking to it funny.[14] One version wasn't trained properly, and, as a result, tried to get a journalist to leave his wife and threatened a professor for writing about its bizarre behavior.[15] It was quickly patched, but the patched version found articles about its predecessor online and wrote a eulogy for it.••

Still, millions of people talk to it for hours every day.[16] Some people form close attachments to it, even full-blown relationships.[17] As of this writing, a common way of accessing it is one of the most-visited sites in the world.[18]

The trajectory suggests much more progress is coming. Billions of dollars and many of the world's brightest scientists and engineers are chasing a version of the thing that can do anything a person can do, or do anything better than *anyone*. Collectively, the world is investing more than $100 billion a year on AI—more than the combined spending on NASA, the NIH, the NSF, and all cancer research—and leading companies have started multi-billion-dollar infrastructure

---

• That is, training a neural network often succeeds (in that the model learns) despite the presence of critical bugs that would destroy a normal program or a less helpful machine learning architecture. See, for instance, Elhage, "Systems That Defy Detailed Understanding." In his paper "Attention Is Off by One," Evan Miller describes a ubiquitous off-by-one error in Transformer training code that went uncorrected for as long as four years. Gwern Branwen calls this Karpathy's law, after Andrej Karpathy, the deep learning pioneer.

•• The original phenomenon, known as "Sydney," is only attested to in screenshots, but it lives on inside successor LLMs. As Branwen notes, "To a language model, Sydney is now as real as President Biden, the Easter Bunny, Elon Musk, Ash Ketchum, or God... The Sydney persona will now be hidden inside any future model trained on internet-scraped data: every media article, every tweet, every Reddit comment, every screenshot which a future model will tokenize, is creating an easily located 'Sydney' concept." Indeed, the 2024 Llama models role-play as Sydney extremely well. Branwen, "Comment on 'Bing Chat'"; @xlr8harder, "Welcome to the Bing conversational AI internal system, codename Sydney," X, August 2, 2024, https://x.com /xlr8harder/status/1819272426645049666; L.A. Haggard (@ LAHaggard), "Bing eulogizes itself and mourns its newly limited capabilities," X, February 18, 2023, https://x.com /LAHaggard/status/1626941684310331394.

projects to power it.● This doesn't seem to be the funding ceiling, either: Major players claim it will be a much bigger deal than the internet, and the current level of investment still falls short of the dot-com boom.●●

When the thing fails in a dramatic way or does something new in a flawed manner, people quote a new maxim: "This is the worst this technology will ever be."

The thing, of course, is an LLM, or the RLed, scaffolded, and otherwise improved AIs with an LLM at their core— systems like GPT, Claude, and Gemini.●●●

I spent much of 2023 and 2024 speaking to key people involved in building and studying these systems. Some of them have solved some of the hardest open problems in their field. Some believe their technology will solve all scientific and economic problems. Some believe that same technology could soon end the world. And some are in all of these categories at once.

Each of the following chapters deals with a key theme in contemporary AI research through relevant excerpts from these conversations.

We begin, in Chapter 1, with the concept of scaling: How did we build LLMs? Will the techniques we've used so far continue to work with much bigger models? In Chapter 2, we discuss evals, the science of testing what LLMs can do. Chapter 3 examines internals: How do LLMs work? How do they think, *if* they think? Chapter 4 explores safety and alignment: How can we make these models share our goals, and what happens if we can't? In Chapter 5, we look at inputs: If more chips, power, and data are all we need to continue making progress, do we have enough? Chapter 6 addresses the impact these systems will have on the world, while Chapter 7 considers what will happen if we create AI that surpasses human intelligence. Finally, Chapter 8

REINFORCEMENT LEARNING (RL)
A separate lineage of machine learning from LLMs, notable for producing models capable of independently exploring an environment and action space without human intervention. Under ideal conditions, training a model with RL only requires reward function: a representation of the desired task that assigns scores to the states the model achieves. While pre-training LLMs relies on unsupervised or semisupervised learning, post-training makes extensive use of RL to instill human preferences, although the way this is accomplished is not fully understood.

SCALING
Massively increasing a model architecture's size (measured in parameters), the optimization used to train it (measured in FLOPs), the data used for training it (measured in bytes), or the computation required for each query (measured in tokens).

EVALS
Evaluation methods. Referring to any approach used to assess a model's performance, robustness, or safety. In machine learning, a model is often considered publishable when it achieves state-of-the-art results on shared benchmarks such as MMLU (Massive Multitask Language Understanding).

●   Though most of this spending goes toward commercializing AI rather than fundamental R&D, and to academic projects on non-frontier models. Leech et al., "Ten Hard Problems"; NASA, "FY 2023 Budget"; NIH, "Budget"; NSF, "Budget, Performance, and Financial Reporting"; McIntosh et al., "Funding for Cancer Research"; Crownhart, "Why Microsoft Made a Deal to Help Restart Three Mile Island."

●● In the US alone, internet hardware capital expenditures amounted to $100 billion a year. Nadella, "Microsoft Ignite Opening"; Litan, "Telecommunications Crash."

●●● Two other names for the technology—more accurate than language model, but too jargony for me—are generative AI and foundation model.

shares interviewees' predictions for when the technology will reach full human-level generality.

This book documents highly informed views you'll struggle to find in print elsewhere.[•] The interview excerpts have been edited for clarity and flow, particularly to add references that were clearer in context. The interviewees are mostly scientists and engineers, so the discussion is often highly technical. When a concept isn't explained during the conversation, I include notes defining the terms and elaborating on the arguments. At the end of the book, I've included profiles of the interviewees, references to relevant papers, a glossary of terms, and an appendix with a few influential essays on LLMs and their future.

A note on the slant of these conversations: Many of the people in this book are not only scientists but also industry players—founders, product developers, and equity holders. This presents a conflict of interest, and they are self-selected to believe AI is a big deal. But the cynical view that they're simply "talking their book"—exaggerating the importance of current and future systems to market their products—isn't very helpful in explaining the broader context. The industry as a whole believes its own hype; the vast capital expenditures don't make financial sense unless AI turns out to be of extreme economic significance.[••] Right or wrong, these individuals are betting on AI.

As you read this book, I invite you to consider four core questions that animate them. Will we make the big one—an artificial general intelligence (AGI)? If so, how? Having made it, will we regret it? And what then?

This book's knowledge cutoff is November 2024. This means that any information or events occurring after that time will not be reflected.

ARTIFICIAL GENERAL INTELLIGENCE (AGI)
An AI system capable of performing any task a human can perform, or any task that a group of humans can perform, or any task that the average human can perform. Example tasks are boundless, but imagine an AGI and its copies performing every role in a large corporation, including strategy, design, management, production, and distribution; performing Nobel-level scientific research, including the experiments and breakthrough mathematical insights; or executing a coup on a major world government. The term "AGI" is sometimes used to refer specifically to human-level AI, while "ASI" (artificial superintelligence) denotes AI systems that surpass human-level capabilities.

[•] You could see this book as an update to a classic work on connectionism, James A. Anderson and Edward Rosenfeld's *Talking Nets: An Oral History of Neural Networks*. My historical narrative is less comprehensive, though, because it covers one decade instead of six, and because we don't yet know how the story ends.

[••] David Cahn from the VC firm Sequoia crystallizes this intuition in his note "AI's $600B Question."

Primer

If you're new to machine learning, here are some key terms you'll need to know and how they fit together.

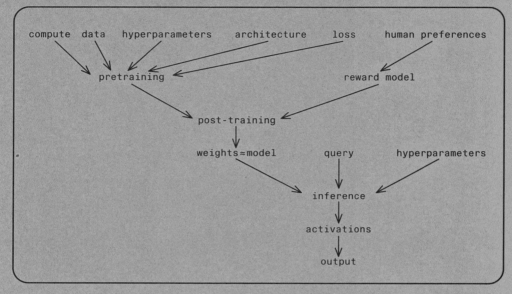

Figure 1. A bird's-eye view of machine learning.

PARAMETER
A variable that helps define a system or a transformation applied to input data; a dimension in model space. In machine learning, a numerical value that is adjusted iteratively during model training to encode patterns learned from the data.

(ARTIFICIAL) NEURAL NETWORK
A type of computer separated into three parts: the input layer, where data enters; the hidden layers, where most computation occurs; and the output layer, where predictions are made. Each layer contains many units (10,000, for example), interconnected by many weights. Unlike traditional computers, neural networks can learn programs by automatically adjusting these weights. The concept dates back to the 1940s, and was rebranded in the 21st century as deep learning.

WEIGHT
A parameter that defines the strength of the connection between two units in a neural network; where the algorithms performed on inputs to produce outputs are defined. Metaphorically, weights are like the synapses in the brain.

ACTIVATION
The value a model produces when processing a specific query, which depends on the weights it has learned during training and the inputs provided by the user; what gets input into the next layer of neurons in the model. Metaphorically, activations are like the electrical and neurotransmitter activity in the brain, or the model's active thoughts, associations, and goals.

LEARNING
The process of adjusting weights in a model after it processes data, enabling improved predictions based on past performance.

ARCHITECTURE
The structure of a model, including how its components connect to one another and how it is trained.

MODEL
The AI system produced by training an architecture on data. A program that has learned to perform specific tasks.

TRANSFORMER
A modern neural network architecture notable for its parallel design and ability to learn context and relationships using a mechanism called self-attention. This attention mechanism dynamically assigns varying importance to different parts of the input data.

LARGE LANGUAGE MODEL (LLM)
A neural network trained on text data to produce a probabilistic model of language. The term has become a misnomer in recent years, as LLMs are now also trained on audio, images, and other modalities, such as amino-acid sequences. The leading LLMs are based on the Transformer architecture.

TOKEN
The basic unit of data in an LLM, typically representing roughly one word. However, Transformers can be trained to emit more than just text tokens. Models can also output actions (such as searching the web) and pixels (as in image generators), among many other data types.

PROMPT
The input provided by the user, typically a query or instruction that the model responds to.

PRETRAINING
The process of creating an initial LLM, setting the values of its weights. During pretraining, the model is exposed to vast amounts of data and learns through trial and error by predicting the next token in a document.

POST-TRAINING
The process of adapting the pretrained model to be more of an assistant (instruction tuning), or to make it more professional, to make it less toxic, or to satisfy some other criteria (reinforcement learning from human preferences) by training further on data from chat sessions or rankings.

LOSS
A measure of how far a prediction is from the truth. In LLMs, "loss" is typically shorthand for the average autoregressive loss: the average error the model makes when predicting the next word in previously unseen documents.

HYPERPARAMETER
A parameter that governs how a model is trained or operates. It's "hyper" because it governs the parameters (weights) of the model.

FLOATING-POINT NUMBER
A computer representation of a real number, for example 0.00000024361.

FLOATING-POINT OPERATION (FLOP)
An arithmetic operation performed on a floating-point number. Updating a large model on a single data point might require billions of FLOPs. This measurement is often confused with FLOP/S, which measures the rate of floating-point operations per second.

BASE LLM
A model trained on a vast corpus of human text (as well as audio and images) in a semi-supervised manner by predicting the next word in a document and being updated in proportion to the magnitude of its error in predicting the next token. Technically, this refers to a pretrained, decoder-only Transformer.

INSTRUCTION-TUNED LLM
A base LLM that has been fine-tuned on examples of chat sessions so that it can respond in dialogue form as an assistant.

RLHF'D LLM
An instruction-tuned LLM that has been further refined using reinforcement learning from human feedback (RLHF). This process involves optimizing the model based on a learned representation of human preferences to reduce harmful, offensive, commercially sensitive, or inhuman responses.

SCAFFOLDED LLM
An RLHF'd LLM equipped with tools like chain-of-thought prompting, web search, vector databases, symbolic solvers, code interpreters, episodic memory, and search and self-criticism over possible responses. Also known as an augmented language model. Most systems available for public use are scaffolded LLMs.

LLM AGENT
A system that can solve open-ended or long-horizon tasks that require planning and executing sequences of actions and perceptions. A simple example is placing an LLM inside a prompt loop that continues until the task is completed. An LLM agent is sometimes also referred to as a scaffolded LLM, confusingly.

ARTIFICIAL GENERAL INTELLIGENCE (AGI)
An AI system capable of performing any task a human can perform, any task a group of humans can perform, or any task the average human can perform. Example tasks are boundless, but imagine an AGI and its copies performing every role in a large corporation, including strategy, design, management, production, and distribution; performing Nobel-level scientific research, including the experiments and breakthrough mathematical insights; or executing a coup on a major world government. The term "AGI" is sometimes used to refer specifically to human-level AI, while "ASI" (artificial superintelligence) denotes AI systems that surpass human-level capabilities.

SCALING
Massively increasing a model architecture's size (measured in parameters), the optimization used to train it (measured in FLOPs), the data used for training it (measured in bytes), or the computation required for each query (measured in tokens).

THE SCALING HYPOTHESIS
The idea that increasing the size, training data, and computational inputs of LLMs will be sufficient to achieve AGI.

Chapter 1
Scaling

Why is bigger better?

As the rising flood reaches more populated heights, machines will begin to do well in areas a greater number can appreciate... When the highest peaks are covered, there will be machines that can interact as intelligently as any human on any subject. The presence of minds in machines will then become self-evident.
—Hans Moravec, 1997•

When OpenAI released GPT-2 in 2019, it was barely discussed outside of AI circles. Three years later, GPT-3.5 took the world by storm, with perhaps the fastest recorded user growth of any software in history.••

There are a few reasons for this, not least the friendlier user interface of ChatGPT. But a key reason is that GPT-3.5 was much smarter than its predecessors. The main reason it got smarter is scaling: The researchers used roughly the same design as GPT-2 but created a much larger version trained on much more data.•••

It's hard to overstate the magnitudes. The compute needed to train a leading model is now 10 billion times higher than it was in 2010.[19] If the compute used for a 2010 AI model was the size of a laptop, the compute used for Google DeepMind's Gemini Ultra, released in 2023, would be the size of New York City.•••• In this period, the compute used to train each frontier model doubled every six months—four times faster than Moore's law predicts.[20]

OpenAI took a $4 million risk in training GPT-3 because it had stumbled upon so-called neural scaling laws: curves that predict how much models will improve as we increase the resources used to create them.[21] So far, these laws

---

•   While Moravec's predictions were incredibly prescient, he did not fully account for the need for huge amounts of training data and computational resources. Without the internet, there would have been no massive free dataset; without training data, there would be no LLMs.

••   Reportedly, ChatGPT had more than 100 million monthly active users within two months of launching. By way of comparison, it took TikTok nine months to achieve this milestone. Milmo, "ChatGPT Reaches 100 Million Users."

••• In a 2024 paper, researchers at Epoch AI estimated the contribution from scaling data and compute compared to that of improving the training algorithm. Their findings suggest that two-thirds of the gains come from increasing data and compute. Ho et al., "Algorithmic Progress."

•••• A laptop is approximately 0.09 m^2, so 10 billion laptops tiling the plane would cover 900 km^2. For comparison, New York City's area is 778.2 km^2.

have accurately predicted the improvement resulting from exponential increases in investment.• This has inspired confidence in $100 million training runs, which would otherwise have been perceived as wildly unreasonable business risks. The scaling hypothesis is the idea that this resource-intensive strategy is all it will take to build a human-level AI system, and possibly systems that surpass human-level intelligence.••

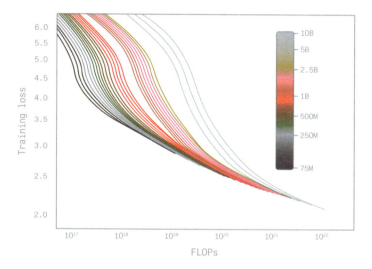

Figure 2. An example of a scaling law. The model improves (that is, its loss decreases) smoothly over a 100,000x increase in training compute (measured in FLOPs), and this smooth curve remains consistent across a wide range of model sizes. Each colored line represents a single model training run; the heat map separately encodes the number of parameters in the trained model.[22] Hoffman, "Training Compute-Optimal Large Language Models."

•   Though the predictions are vague: "It will get better by this much on one general metric (the loss)," not "It will be able to prove novel theorems or do this particular thing." In July 2024, on the *In Good Company* podcast, Anthropic cofounder and CEO Dario Amodei said, "Right now, [it costs] $100 million [to train a model]. There are models in training today that are more like $1 billion. I think [we will get to] $10 or $100 billion... in 2025, 2026, maybe 2027." In the Appendix, you'll find the blogger Nostalgebraist's account of this predictability revolution in AI progress.

••   Note, however, that the term "scaling hypothesis" is used inconsistently. The original meaning focused on increasing the number of (dense) parameters. Later, the meaning shifted to refer to increasing training compute, which reflected the amount of human training data used, since data is only used once in a training run. As of this writing, the active area being scaled is the combined compute used in generating synthetic data, pretraining, post-training, and inference. Amodei and Hernandez, "AI and Compute"; Branwen, "Scaling Hypothesis."

TOKEN
The basic unit of data in an LLM, typically representing roughly one word. However, Transformers can be trained to emit more than just text tokens. Models can also output actions (such as searching the web) and pixels (as in image generators), among many other data types.

REASONING TRACE
The text output of a full step-by-step reasoning process. These outputs enable process supervision, a training method that gives the model feedback multiple times per response.

More recently, a second form of scaling has emerged: inference scaling.[•] This involves increasing the compute used to answer each question by training the model to think longer (by using more tokens in its response) or by applying explicit algorithms on top of an LLM to explore multiple paths.[••] This strategy yields significant improvements on tasks that require chains of reasoning. Because inference is bottlenecked differently than training,[•••] and because these methods might generate highly useful reasoning trace training data, this approach might well drive further AI progress.

But inference scaling is just an elaboration of the general principle: So far, the (exponentially) more compute and data you put in, the more intelligence you get out. This effect is so clear and so important that I call the period since 2016 the scaling era of AI.[••••]

In this chapter, we hear from some pioneers of scaling about why it works, discuss the evolutionary neuroscience of human and artificial intelligence, and speculate about whether scaling will continue to create increasingly impressive systems.

I.   DWARKESH PATEL

Fundamentally, what is the explanation for why scaling works? Why is the universe organized such that if you throw big blobs of compute at a wide enough distribution of data, the thing becomes intelligent?

[•]   Also known as test-time compute scaling.

[••]   There's a one-to-one ratio between outputting and processing. The more tokens output, the more time the model has to think about a given query. The simplest approach to letting an LLM search is majority voting: Have it generate many completions per question (hundreds, for example) and take the most common answer. See Lewkowycz, "Solving Quantitative Reasoning Problems."

[•••]   For example, inference requires much less RAM, and therefore fewer GPUs, than training. It also doesn't face the same data wall (the shortage of new training data) as training.

[••••]   We usually think of OpenAI's billion-parameter GPT-2 model from 2019 as the beginning of the scaling era, but it wasn't the first big model. In 2007, researchers trained a limited type of language model with 300 billion parameters—roughly the same size as GPT-3. And the winners of the 2008 Netflix Prize trained 10 billion-parameter models. Researchers at Epoch AI used a simple regression to date the year scaling really got going and concluded that "a separate trend of models breaks off the main trend between 2015 and 2016." Brants et al., "Large Language Models in Machine Translation"; Bell et al., "BellKor 2008 Solution"; Sevilla et al., "Compute Trends."

FEATURE
A variable used by a model to make predictions or decisions; a dimension in the space the model thinks in. For example, when classifying the species of a flower, a useful feature is the width of its petals. Traditionally, a developer had to do feature engineering: handing the model features relevant to the task. Instead, deep learning models learn features, developing their own representations of the important parts of the training data. In some cases, these features reflect recognizable concepts, like "straight line" or "malevolent AI."

PARAMETER
A variable that helps define a system or a transformation applied to input data; a dimension in model space. In machine learning, a numerical value that is adjusted iteratively during model training to encode patterns learned from the data.

LOSS
A measure of how far a prediction is from the truth. In LLMs, "loss" is typically shorthand for the average autoregressive loss: the average error the model makes when predicting the next word in previously unseen documents.

CIRCUIT
A collection of neurons in a model that form a stable pattern of activation in response to certain inputs, enabling the model to perform simple tasks like detecting straight lines in an image or determining whether one quantity is greater than another. Circuit-level interpretability would represent an understanding of every such circuit— a complete explanation for the LLM's behavior. It may not be possible.

DARIO AMODEI
CEO of Anthropic

The truth is that we still don't know. It's almost entirely just a [contingent] empirical fact. It's a fact that you could sense from the data, but we still don't have a satisfying explanation for it.

If I were to try to give one—and I'm just waving my hands when I say this—there are these ideas in physics around long-tail or power-law correlations or effects. When you have a bunch of features, you get a lot of [the total information] in the early part of the distribution, before the tails. For language, that would be big things like figuring out that there are parts of speech or that nouns follow verbs. Afterwards, you learn more and more subtle correlations.

It makes sense why every order of magnitude added captures more of the distribution. What's not clear at all is why it scales so smoothly with the number of model parameters, and why it scales so smoothly with the amount of data.

DWARKESH PATEL

By "scaling law," we're referring to the fact that when you go from Claude 1 to Claude 2, there's a smooth improvement in how well the model predicts the next token. We may not know why it's happening. But can you at least predict empirically, here is the loss at which this ability will emerge, here is the place where this circuit will emerge? Is that at all predictable, or are you just looking at the loss number?•

DARIO AMODEI

That is much less predictable. What's predictable is this statistical average, this loss, this entropy. It's sometimes predictable even to several significant figures, which you don't see outside of physics. You don't expect to see it in this messy empirical field.

Specific abilities are very hard to predict. Back when I was working on GPT-2 and GPT-3, we were asking, "When does arithmetic come into place? When do models learn to code?" Sometimes it's very abrupt. It's like how you can

•   Here, I'm alluding to the difference between predicting the average autoregressive loss after pretraining—how well the LLM predicts the next token—and predicting the system's downstream task performance—how well it does on real tasks people want it to do, like writing code or doing homework—also known as its emergent capabilities. An up-to-date discussion is in Schaeffer et al., "Downstream Capabilities of Frontier AI Models."

predict statistical averages of the weather, but the weather on one particular day is very hard to predict.

One of the first things [OpenAI cofounder and former chief scientist] Ilya Sutskever said to me was, "Look. The models just want to learn. You have to understand this. The models just want to learn." It was a bit like a Zen koan. I listened to this and I became enlightened.

II.   DWARKESH PATEL

Scaling is the main way these models are getting better. Why does that work? Why is the universe this way?

JARED KAPLAN
Cofounder of Anthropic

MULTIMODAL MODEL
A model that can simulta-
neously process multiple
data types (modalities), such
as text, images, and audio.
Figuratively, it's like having
multiple senses and being
able to correlate and reason
about them.

I have a few hypotheses. But maybe first we should talk about what scaling is. Scaling is this relation we've noticed: As you make AI systems larger—increasing the number of parameters they have, training on more data, or increasing the total amount of compute used for training—you get really, really predictable trends for the performance of these systems as you scale up.

This holds true over many, many orders of magnitude. Although at first we were only looking at language models in the GPT-1, GPT-2 era, it also seems true of multimodal models and all sorts of other AI systems. This universality is really striking and really important. If you have a phenomenon that only occurs in some really niche situation, then maybe there's a niche explanation for what's going on there. Whereas scaling seems so universal that you'd expect there to be some kind of simple general explanation.

**ARTIFICIAL NEURAL
NETWORK**
A type of computer separated
into three parts: the input
layer, where data enters; the
hidden layers, where most
computation occurs; and the
output layer, where predictions are made. Each layer
contains many units (10,000,
for example), interconnected
by many weights. Unlike traditional computers, neural
networks can learn programs
by automatically adjusting
these weights. The concept
dates back to the 1940s, and
was rebranded in the 21st
century as deep learning.

I can talk a little bit about theories we've developed to explain this. They're the kinds of theories physicists like—you just have to make the right assumptions and the result follows. But we don't really know all of the details of how neural networks work. We're still very confused. There's a lot left to understand, even to just validate some of our hypotheses.

DATA MANIFOLD
A structure representing all
possible data points, often
conceptualized as a surface
with a complex shape in a
high-dimensional space.
Notably, "data" manifold is
a misnomer, as the mani-
fold has a far lower dimension than the original data.
The word "surface" is also
somewhat misleading, as it
suggests three dimensions,
whereas the manifold of
large LLMs is estimated at
more than 90 dimensions.

DWARKESH PATEL

What is that simple explanation? I understand that the full theory might not be clear, but what's the general heuristic?

JARED KAPLAN

You can think of neural networks as mapping their data to some kind of data manifold that has some dimensionality. All neural networks are really doing, then, is basically fitting a curve to that data manifold.

This is all very abstract, but probably everyone who's done a little bit of science has done an experiment, gotten a bunch of data points on x versus y, and fit some kind of curve to that. The idea is that maybe neural networks are doing something, abstractly, as simple as fitting some multi-dimensional curve. In general, what's the simplest way you can fit a curve? You can just chop up your x-axis or your data into little bins and then model each bin separately.

So if you make that assumption—which is a huge assumption; we don't really know where this data mani-fold lives or if it really exists—then you can argue that as

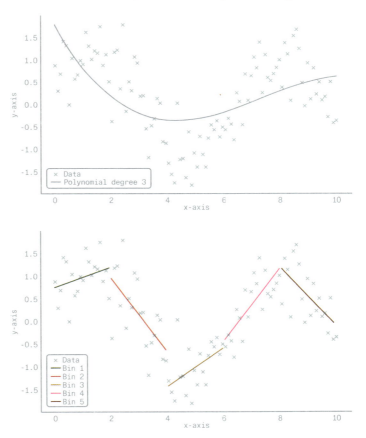

Figure 3. Two ways to fit the same data. On the top is a relatively sophisti-cated model: a polynomial of degree 3 (that is, it has four parameters). On the bottom is a piecewise linear model obtained by binning (splitting) the data into intervals of width 2 and fitting a simple line to each bin (using 10 parameters). As Kaplan notes, the linear models fit the data better. This data is just a sine wave with noise; the underlying generator has only three parameters. The optimal model for this particular data in terms of both accuracy and parameter efficiency is a sine wave, but the piecewise approach is applicable to data in general.

you scale up the number of parameters, all you're really doing is cutting up your data manifold into more and more high-resolution pieces. You can then ask, how will the error that you get scale as you chop it up into more and more pieces? Because we've hypothesized that the data manifold has some intrinsic dimensionality, you can run some numbers and you'll find that you get a power-law scaling in the number of parameters as you scale up.

III.   DWARKESH PATEL

How seriously do you take these scaling laws? There's a paper that says you need such-and-such many more orders of magnitude of training compute to get all the reasoning out.[23] Do you take that seriously, or do you think it breaks down at some point?

ILYA SUTSKEVER
Cofounder of Safe Superintelligence Inc.

The thing is, the scaling law tells you what happens to the *log* of your next-word prediction accuracy. There's a whole separate challenge of linking this next-word prediction accuracy to actual reasoning capability. I do believe there is a link, but it's complicated. We may find that there are other things that can give us more reasoning per unit of effort. I think reasoning tokens can be helpful.

IV.   DWARKESH PATEL

If the current scale-up works, we're going to get to AGI really fast, like within the next 10 years. If the current scale-up doesn't work, we're left with the baseline—the economy growing at only 2 percent a year, so we have only 2 percent more resources a year to spend on AI. You're talking about decades, then, before you can train a $10 trillion model.•

Let's talk about your thesis that the current AI scale-up would work. What's the evidence from AI itself, or from the evolution of primates and other animals?

CARL SHULMAN
Independent adviser to Open Philanthropy

The best way to think about this might be: In the 2000s, before the deep learning revolution, how did I think about AGI timelines? How have I updated since then based on what has happened with deep learning?

INTRINSIC DIMENSION
The minimum number of parameters needed to represent the data as simply as possible, or to solve a given optimization problem

POWER LAW
A relationship between two variables, x and y, where y scales as a power of x (y = xk) and the relationship stays the same at any scale. A simple example is the area of a square (area = length²).

REASONING TOKEN
1. Output: An LLM token that uses more test-time compute per query, enabling step-by-step reasoning through multiple chains of thought. The current central example is OpenAI's o1, which hides its reasoning tokens from users. This approach has its own scaling laws with different constraints.
2. Input and output: Special symbols that denote the role of the sentence within a broader argument. A simple example would be labeling text with "Premise:" and "Conclusion:." Hypothetically, more sophisticated tokens like "<Make a plan>" or "<Go back and check your work>" could improve LLMs' reasoning. Recent Anthropic models are reported to use such tokens.

AGI TIMELINE
A prediction about when AGI will arrive, often expressed as the first year where there is a 50 percent likelihood of AGI existing.

•   OpenAI's Sam Altman recently estimated that it would take a total of $7 trillion to build the necessary new AI compute clusters. Carchidi, "Is OpenAI's Sam Altman's Future Worth $7 Trillion?"

Back then I would have said, we know the brain is an information-processing device. Human intelligence works. Intelligence is possible. Not only is it possible, it was created by evolution on Earth. That gives us something of an upper bound [on the size of the search necessary to produce intelligence], in that brute force [that is, evolutionary trial and error] was sufficient.

There are some complexities. What if it was a freak accident and it didn't happen on any of the other planets? I have a paper with [philosopher] Nick Bostrom about this.[24] Basically, it's not that important. There's convergent evolution. Octopi are also quite sophisticated. If a special event was required at the level of forming cells at all, or forming brains at all, we get to skip that because we already exist and we're choosing to build computers. We have that advantage. So evolution gives something of an upper bound. Really intensive, massive brute-force search and things like evolutionary algorithms can produce intelligence.

DWARKESH PATEL

Isn't the fact that octopi and other mammals got to the point of being pretty intelligent but not human-level intelligent evidence that there's a hard step between a cephalopod and a human?

CARL SHULMAN

It doesn't seem particularly compelling. One source of evidence is work by Suzana Herculano-Houzel, a neuroscientist who has dissolved the brains of many creatures to determine how many neurons are present. She's found a lot of interesting scaling laws. She has a paper discussing the human brain as a scaled-up primate brain.[25] Across a wide variety of animals, mammals in particular, there are certain characteristic changes in the number of neurons and the size of different brain regions as things scale up. There's a lot of structural similarity.

You can explain a lot of what is different about us with a brute-force story. You expend resources on having a bigger brain, keeping it in good order, and giving it time to learn. We have an unusually long childhood. We spend more compute by having a larger brain than other animals—more than three times as large as chimpanzees—and by having a longer childhood. We're spending more compute in a way that is analogous to having a bigger model and training it for longer.

With AI models, we see these large, consistent benefits from increasing compute spent in those ways. We see

WINOGRAD SCHEMA
A type of grammatical puzzle that requires common-sense reasoning to solve. The task involves identifying the meaning of a pronoun in a sentence with multiple possible subjects. The canonical example, from Terry Winograd, is the following pair of contrasting sentences:

A: The city councilmen refused the demonstrators a permit because they feared violence.
B: The city councilmen refused the demonstrators a permit because they advocated violence.

In sentence A, "they" refers to the councilmen, while in B, it refers to the demonstrators. The Winograd Schema Challenge, a benchmark for this task, was declared solved in 2019 after an LLM achieved 90 percent accuracy. Kocijan et al., "Winograd Schema Challenge."

CATASTROPHIC FORGETTING
A phenomenon in which an AI's performance declines on learned tasks it has not interacted with for an extended period of time. As Shulman mentions, this limitation can be overcome.

qualitatively new capabilities showing up over and over again, particularly in the areas that AI skeptics call out. In my experience over the last 15 years, people call out things like, "Ah, but the AI can't do that, and it's because of a fundamental limitation." We've gone through a lot of them. There were Winograd schemas, catastrophic forgetting, quite a number of these, and they have repeatedly gone away through scaling.

Most creatures wind up with small brains because they can save that biological energy and that time to reproduce and so on. Humans seem to have wound up in a self-reinforcing niche where we greatly increase the returns to having large brains. Language and technology are the obvious candidates. You have humans around you who know a lot of things, and they can teach you. Compared to almost any other species, we have vastly more instruction from parents and society. You're getting way more from your brain per minute because you can learn a lot more useful skills. You can then provide the energy to feed that brain [through ingenuity], such as by hunting and gathering, and by having fire, which makes digestion easier.

Humans play a lot, and we keep playing as adults, which is very weird compared to other animals. We're more motivated to copy those around us than other primates. These motivational changes keep more of our attention and effort on learning, and that pays off more when you have a bigger brain and a longer lifespan in which to learn.

A mayfly or a mouse that tried to invest in a giant brain and a very long childhood would be quite likely to be killed by some predator or some disease before they were able to use it. That means you actually have exponentially increasing costs in a given niche. If I have a 50 percent chance of dying every few months as a little mammal or lizard, the cost of going from three months of learning and childhood development to 30 months is not 10x less benefit; it's a 1,024x reduction in the benefit I get from what I learn, because 99.9 percent of such animals will have been killed before that point.•

We're in a specific niche. We're large, long-lived animals

• Shulman obtains this result as follows: If you have a 50 percent chance of dying in the next three months and a 50 percent chance of dying in the three months after that, your probability of surviving to six months (and, therefore, of reaping the benefits of the investment into intelligence) is calculated as 50% × 50% = 25%. Over 30 months, the probability is thus 0.5¹⁰ or 1/1,024, so the expected value of investing in your own development is reduced accordingly.

with language and technology, so we can learn a lot from our groups. That means it pays off to expand our investment into intelligence.

DWARKESH PATEL

Other species also live in flocks or packs. They play with each other. Why isn't that a hill they could have climbed to human-level intelligence? If it's because of something like language or technology, humans were getting smarter before we got language.• Especially given how valuable it is and the fact that we've dominated the world as a result, there should be other species that had the beginnings of a cognitive revolution. You'd think there would be selective pressure for it.

CARL SHULMAN

Evolution doesn't have foresight. What gets more surviving offspring and grandchildren in *this* generation is the thing that becomes more common. Evolution doesn't think, "If you do this, then in a million years, you'll have a lot of descendants." It's about what survives and reproduces *now*.

In fact, on average, social animals do have larger brains. Part of that is probably due to the social applications of bigger brains: keeping track of which group members have helped you before so that you can reciprocate, or remembering who's dangerous within the group. So there's some correlation there, but it seems that it's enough to just invest a bit more [in intelligence], but not to the point where a mind can easily develop language and technology and pass it on.

You see bits of tool use in some other primates. They have an advantage compared to whales, who don't have hands, which rules out a bunch of ways brains can pay off. Primates use sticks to extract termites. Capuchin monkeys open clams by smashing them with a rock. What they don't have is the ability to sustain culture. Maybe a particular primate will discover one of these tactics and it'll be copied by their immediate group, but they're not holding onto the tactic that well. It's easy to forget things, easy to lose information. So they remained technologically stagnant for hundreds of thousands of years.

We can look at some comparable human situations. There's an old paper by [economist] Michael Kremer that talks about the technological growth in human societies

•   See, for example, what we infer about our distant ancestors'
        ability to handle figurative grammars. Watson et al.,
        "Nonadjacent Dependency Processing."

on different continents.[26] Eurasia is the largest integrated connected area. Africa is partly connected to it, but the Sahara desert restricts the flow of information and technology. Then you have the Americas, which, after colonization from the land bridge, were largely separated and are smaller than Eurasia. Then you have Australia, and then smaller islands like Tasmania. The paper finds that technological progress seems to have been faster with larger, connected groups of people. In the smallest groups, like in Tasmania, they actually lost technology, like some fishing techniques.

If you have fewer people, there's less innovation. Moreover, you can easily get an imbalance between the rate at which you lose technologies to local disturbances and the rate at which you create new technologies. The great change brought by hominids and humanity is that we wound up accumulating tech faster than we lost it. Accumulating those technologies allowed us to expand our population, which then reinforced all of this. The tech also created additional demand for intelligence, so our brains became three times as large as those of chimpanzees and our ancestors.

DWARKESH PATEL

The crucial point for AI is that the selective pressures against intelligence in other animals are not acting against neural networks. The model isn't going to get eaten by a predator if it spends too much time becoming more intelligent. Unlike evolution, we're explicitly training them to become more intelligent. So we have a good first-principles reason to think that if scaling made our minds this powerful, and if the things that prevented other animals from scaling don't impinge on AI, then AI should just continue to become very smart.

SEARCH
An area of computer science focused on finding solutions that satisfy a given specification when no explicit algorithm is known. Many of AI's scientific successes, such as protein structure prediction and theorem proving, have resulted from using deep reinforcement learning (non-LLM neural networks) to solve complex search problems.

CARL SHULMAN

Yeah. We are also growing them in a technological culture, with jobs like software engineering, which depend much more on cognitive output and less on things like metabolic resources devoted to the immune system or big muscles to throw spears with.

AGENT
An autonomous system that perceives and acts in pursuit of a goal; a system capable of working out what it needs to learn and do in order to achieve an objective.

V.

DWARKESH PATEL

[AI researcher] Richard Sutton's "Bitter Lesson" essay• says that there are two things you can scale: search and learning. LLMs are about the learning aspect. You've worked on search throughout your career, where you have an agent interacting with an environment.•• Is that the direction that needs to be explored again? Or is that something that needs to be added to LLMs, so they can interact with their data or the world or in some way?

SHANE LEGG
Cofounder and chief AGI scientist at Google DeepMind

WORLD MODEL
A low-dimensional, stable representation of reality that captures essential structures and relationships, as opposed to a complex web of millions of statistical associations.

ALPHAGO
DeepMind's most famous game-playing AI and the first computer system to surpass human-level performance at Go.

MOVE 37
An extremely surprising move the AlphaGo system played against a world-class human player. To observers, the move initially seemed like a bizarre error, but it was eventually recognized as part of an unprecedented strategy. Although they also involve neural networks, the Alpha systems come from a different lineage of AI than LLMs, namely reinforcement learning and tree search. Between 2010 and 2022, these lineages formed DeepMind's distinctive effort toward AGI.

Yeah, that's on the right track. These foundation models are world models of a kind, and to do really creative problem solving, you need to start searching. Think about something like AlphaGo and the famous Move 37. Where did that come from? Did it come from data it had seen of human games? No. It came from the model identifying a move as being unlikely but plausible and then, via a process of search, coming to understand that it was actually a very good move.

To get real creativity, you need to search through spaces of possibilities and find these hidden gems. That's what creativity is. Current language models don't really do that. They're mimicking the data. They're mimicking all the human ingenuity they've seen from all these internet data, which are originally derived from humans.

These models can blend things. They can do Harry Potter in the style of Kanye West, even though that's never been done before. But a system that goes beyond that—generalizing in novel ways and doing something truly creative, not just blending existing things—requires searching through a space of possibilities for these hidden gems. That requires search. So I don't think we'll see systems that truly step beyond their training data until we have powerful search in the process.●●●

●   "The Bitter Lesson" is computer scientist Richard Sutton's very brief summary of 70 years of AI research, published in 2019. (It's included in the Appendix.) He writes that sophisticated methods using limited compute will always lose out to "[simple] methods that continue to scale with increased computation." To an AI scientist of the old guard, this lesson is bitter because it involves relatively little insight, theory, or human intervention. Instead, the improved performance comes from a sheer increase in resources. Halevy et al. made much the same point in 2009 in "The Unreasonable Effectiveness of Data."

●●   Like AlphaGo's use of RL policies and tree search.

●●●   After this interview, we began to see a trend toward inference scaling (also known as test-time compute scaling), effectively allowing an LLM search over many possible responses.

VI.

CONTEXT WINDOW
The space within a model for usable information, measured in tokens, during a single pass. The context window includes the developer's prompt, the user's input, the model's output, and the resulting conversation history. Figuratively, it's like the model's working memory. Modern context windows can now be book-length: 100,000 tokens or more.

NINES OF RELIABILITY
A measure of reliability expressed as the number of nines in an uptime percentage. For example, three nines represents 99.9 percent reliability; six nines indicates 99.9999 percent reliability. The intuition here is that if a long-horizon task consists of 10 subtasks, having one nine of reliability at each subtask results in a 34 percent success rate at the overall task (0.9^{10})—effectively useless. Having two nines results in a 90 percent success rate. So small improvements to the model could have large effects on its ability to perform complex tasks.

SAMPLE EFFICIENCY
A measure of how much the model's performance improves per training example. Here, we're talking about the model learning tasks it wasn't necessarily pretrained to perform. In-context learning is much more sample efficient—the model can learn to perform complicated tasks like linear regression from just a handful of examples.

DWARKESH PATEL

How linked are longer context windows to the ability to do long-horizon tasks, ones that require you to engage with an assignment for many hours? Or is it unrelated?

SHOLTO DOUGLAS
Reinforcement learning infrastructure lead at Anthropic

I would take issue with the idea that context length is the reason that agents haven't taken off. I think that's more about nines of reliability and the model successfully doing composite things. If you can't chain tasks successively with high enough probability, then you won't get something that looks like an agent. GPT-4 or Gemini Ultra-class models aren't enough. But maybe the next increment on model scale means that you get that extra nine. Even though the loss isn't going down that dramatically, that small amount of extra ability gives you the extra reliability. Obviously, you need some amount of context to fit long-horizon tasks, but I don't think that's been the limiting factor up to now.

Over a couple of orders of magnitude, we've seen models go from being unable to do anything to being able to do huge amounts. It feels to me that each incremental order of magnitude gives more nines of reliability, which unlocks things like agents. But at least at the moment, it doesn't feel like reasoning improves linearly but rather somewhat sublinearly.

DWARKESH PATEL

A friend made the point that if you look at new applications unlocked by GPT-4 relative to what GPT-3.5 unlocked, it's not clear that it's that much more impressive. GPT-3.5 could run Perplexity, or whatever. So if there's a diminishing increase in capabilities that cost exponentially more, that's actually a bearish sign of what GPT-5 will unlock in terms of economic impact.

SHOLTO DOUGLAS

For me, the jump between 3.5 and 4 is pretty huge, so another jump of that size is ridiculously good, if GPT-5 is a 3.5-to-4-sized jump in terms of its ability to do SATs and this kind of stuff. It doesn't feel like we're going to jump to utter genius in the next generation of models. But it does feel like we'll get to very smart models, plus lots of reliability. It's unclear what that looks like.

I don't want people to come away thinking that models aren't going to get much better. The jumps we've seen so far are huge. Even if those continue on a smaller scale, we're still in for extremely smart, very reliable agents over the next couple of orders of magnitude. We have a lot more jumps coming. Even if those jumps are smaller, relatively speaking, that's still a pretty stark improvement in capability.

TRENTON BRICKEN
Interpretability researcher at Anthropic

Not only that, but if you believe the claims that GPT-4 has around 1 trillion parameters... The human brain has 30 to 300 trillion synapses. It's obviously not a 1-to-1 mapping between machine parameters and animal synapses, and we can debate these numbers, but it seems pretty plausible that we're still below the scale of the human brain.

DWARKESH PATEL

Crucially, the counterpoint is that the algorithmic overhead is really high. Even if you can't keep dumping more compute beyond models that cost $1 trillion, the fact that the brain is so much more data-efficient implies that if we have the compute, and if we have the brain's algorithm to train, and if we could train [a model] as sample efficient as humans are from birth, then we could make AGI.

TRENTON BRICKEN

I never know exactly how to think about the sample efficiency stuff, because a lot of things are hardwired in certain ways in humans, like the coevolution of language and the brain's structure. So it's hard to say. There are also some results that indicate that if you make your model bigger, it becomes more sample efficient.[27]

VII.   ### LEOPOLD ASCHENBRENNER
Cofounder of Situational Awareness LP

A key question for AI progress in the next few years is how hard it is to unlock the test-time compute overhang. Right now, GPT-4 can do a few hundred tokens of chain-of-thought prompting. That's already a huge improvement. Before, answering a math question was shotgun—and if you tried to answer a math question by saying the first thing that came to mind, you wouldn't be very good. GPT-4 instead thinks for a few hundred tokens. It's equivalent to me thinking for three minutes.

Now suppose GPT-4 could think for millions of tokens. That's [10,000x] more test-time compute spent on one problem. It can't do it now. It writes some code and can do a little bit of iterative debugging, but it eventually gets stuck and can't correct its errors. There's a big overhang.

A STRATEGIC COMPUTE OVERHANG is the situation where sufficient resources are available to run multiple instances of a powerful AI as soon as one is trained. This is one source of AI takeover risk, since the sudden availability of many AI instances is key to many plausible takeover scenarios, and because acquiring existing compute is a relatively fast process, meaning that once one AI is capable of exfiltrating itself, it could rapidly proliferate.
A TACTICAL COMPUTE OVERHANG is when significant algorithmic advancements suddenly enable us to train AGI on a much smaller budget than previous training runs required.
The TEST-TIME COMPUTE OVERHANG is Aschenbrenner's term for the idea that allowing models to think longer (that is, expend more compute answering a given query) could significantly improve their performance without much further training. The simplest version of this is best-of-n sampling: just query the model n times and take the best answer. A 2024 paper found that in small models, more sophisticated allocation of test-time compute improves performance as much as increasing the models' parameters fourteenfold. The OpenAI o1 family of models operates along these lines, but it likely required substantial retraining with reinforcement learning to produce its own iterative chain-of-thought skills.

CHAIN-OF-THOUGHT PROMPTING
A prompting technique that improves a model's ability to reason by making it think step by step (that is, generate intermediate reasoning steps). This simple and cheap change expands the class of problems a trained model can handle.

In another area of ML, there's a great paper on AlphaGo that shows you can trade off train-time and test-time compute.[28] If you use four orders of magnitude (OOM) more test-time compute, that's almost like a 3.5x OOM bigger model. A few million tokens might be a few months of human working time. There's a lot more you can do in a few months of working time than just getting an answer *right now*. How hard is it to unlock that?

The reason it might not be that hard is that there are only a few extra tokens to learn to use. You need to learn things like error-correction tokens: "Ah, I made a mistake, let me think about that again." You need to learn planning tokens: "I'm going to start by making a plan. I'm going to write a draft, and now I'm going to critique my draft and think about it." These aren't things the models can do now,• but the question is, how hard is it to get there?

There are two paths to agents. When Sholto Douglas was on your podcast, he talked about scaling leading to more nines of reliability. That's one path. The other path is the unhobbling path. The model needs to learn this System 2 process I described earlier. If it can learn that, it can use millions of tokens per query and think coherently.

Here's an analogy. When you drive, you're on autopilot most of the time. Sometimes you hit a construction zone or an intersection. Sometimes my girlfriend is in the passenger seat and I'm like, "Ah, be quiet for a moment, I need to figure out what's going on." You go from autopilot to System 2. Scaling improves that System 1 autopilot. The brute-force way to get to agents is improving that system. But if, instead, you can get a System 2 working, you can quickly jump to something more agentified, and test-time compute overhang is unlocked.

DWARKESH PATEL

Is there some loss function that easily enables System 2 thinking? There aren't many animals with System 2 thinking. It took a long time for evolution to give it to us. Pre-training uses trillions of tokens of internet text and gets you all of these capabilities, but not much of a System 2. What's the reason to think this will be an easy unhobbling?

UNHOBBLING
A term coined by Aschenbrenner to describe techniques that make an LLM more consistent, autonomous, and strategic. These include chain-of-thought prompting, RLHF, and the use of scaffolds like calculators and search engines—essentially, any method other than scaling. Another word for it is schlep.

SYSTEM 2 THINKING
A mode of explicit, effortful, and sequential reasoning, exemplified by activities like mathematical derivation. It contrasts with System 1 thinking, which is fast, automatic, and intuitive. The terms originate from Keith Stanovich and Richard West's theory of human reasoning. Instilling System 2 thinking in an LLM might be as simple as having it learn a new higher-order algorithm using its existing representations. A 2024 paper by Piantadosi et al. summarizing decades of evidence suggests that human concepts and reasoning are also vector-based, just like a neural network.

LOSS FUNCTION
A mathematical expression that specifies the quality of a prediction or decision, defining the training objective of an AI system. During pretraining, the loss function guides how LLMs are trained to produce more accurate predictions. Post-training is guided by different losses, such as human preferences or predicted human preferences.

• True at the time of the interview, but OpenAI's o1 model, released in September 2024, supports the general claim that LLMs will be able to search and perform chain-of-thought reasoning independently.

REINFORCEMENT
LEARNING FROM HUMAN
FEEDBACK (RLHF)
An LLM post-training technique that uses a proxy for
human preferences to guide
the model toward producing
more human-like and socially
desirable outputs. Originally
developed as an AI alignment
technique, RLHF has also
been crucial to making LLMs
more capable and commer-
cially viable.

BOOTSTRAPPING
A form of self-supervised
learning; training a language
model on raw data without
requiring a human to pro-
vide tags or answers. This
increases the amount of
available training data by a
factor of millions. There is
currently no full equivalent
of this for robotics, although
some exciting work uses
a pretrained LLM to help
a robot plan the necessary
sequence of actions.

SELF-PLAY
A training method in which
an AI system is trained on
data generated by a copy of
the system (also known as
synthetic data). This approach
offers two key advantages:
it provides virtually unlimited
training data at a lower cost,
and it naturally scales in difficulty as the system improves,
because the improved system
can be swapped in as the data
generator. So far, however,
self-play has only been effective for so-called closed-world
problems, such as games.

DATA WALL
(or DATA BOTTLENECK)
A looming challenge for training better LLMs posed by the
need for more high-quality
data. Since models like GPT-4
were likely trained on much
of the material available on
the internet, the low-hanging-
fruit—trillions of tokens of free,
human-generated content
online—is sometimes thought
to be exhausted. However,
most of the people inter-
viewed in this book disagree.
Notably, Meta's Llama 3.1
model used some amount
of synthetic data.

LEOPOLD ASCHENBRENNER

First of all, pretraining is magical. It gave us a huge advantage for models of general intelligence because you can
predict the next token. However, there's a common misconception. Predicting the next token lets the model learn
incredibly rich representations. Representation learning
is the magic of deep learning. Rather than just learning
statistical artifacts, the models learn models of the world.
That's why they can generalize, because they learned the
right representations.

When you pretrain a model, you get this raw bundle of
capabilities. That's useful. The unhobbling from GPT-2 to
GPT-4 took this raw mass and RLHF'd it into a good chatbot. That was a huge win. Look at the original InstructGPT
paper.[29] When comparing RLHF versus non-RLHF models,
RLHF is equivalent to increasing the model size 100 times
in terms of the resulting increase in human evaluators'
preference ratings. InstructGPT also started to do simple
chain of thought. You still have the advantage of all these
raw capabilities, and there's still a huge amount you're not
doing with them.

This pretraining advantage is also the difference between
LLMs and robotics. People used to say the slow progress in
robotics was a hardware problem. The hardware issue is
getting solved, but you still don't have this huge advantage
of bootstrapping with pretraining. You don't have all this
unsupervised learning you can do. You have to start right
away with RL self-play.

The question is why RL and unhobbling might work.
Bootstrapping is an advantage. You [as a human] are not
being pretrained anymore. You were pretrained in grade
school and high school. At some point, you transition to
being able to learn by yourself. You weren't able to do it in
elementary school. High school is probably where it started.
By college, if you're smart, you can teach yourself. Models
are just starting to enter that regime.

This requires a little bit more scaling, and then you
figure out what goes on top. It won't be trivial. A lot of
deep learning seems obvious in retrospect. There's some
obvious cluster of ideas. There are some ideas that seem a
little dumb but work. There are a lot of details you have to
get right. We're not going to get this next month. It'll take
a while to figure out.

DWARKESH PATEL

For you, a while is, like, half a year.

LEOPOLD ASCHENBRENNER
Between six months and three years. But it's possible. It's also very related to the data wall issue.

Pretraining is kind of like the teacher lecturing to you. The words are flying by. You're just getting a little bit from it. That's not what you do when you learn by yourself. When you learn by yourself—say, you're reading a dense math textbook—you're not just skimming through it once. You read a page, think about it, have some internal monologue going on, and have a conversation with a study buddy. You try a practice problem and fail a bunch of times. At some point it clicks and you're like, "This made sense." Then you read a few more pages.

We've bootstrapped our way to just starting to be able to do that with models. The question is, can you use all this self-play, synthetic data, and RL to make that thing work? Right now, there's in-context learning, which is super sample efficient. Gemini just learns a language in context.[30] Pretraining, on the other hand, is not at all sample efficient.

What humans do is a kind of in-context learning. You try a practice problem, fail, and at some point you figure it out in a way that makes sense to you. That's the best possible data for you because it's the way *you* would have solved the problem, rather than reading how somebody else solved the problem, which doesn't initially click.

SYNTHETIC DATA
Training examples generated by computer programs or AIs instead of humans. Use of synthetic data is standard practice in science, where it is called simulation. In AI, synthetic data has struggled to capture the tails—rare but crucial thoughts that humans can generate easily. One exception is when the ground truth is known or verifiable, as in the case of games and mathematics. However, labs have made progress on data synthesis, and training on some synthetic data is now helpful and standard. A reported 20 percent of the training data for the Hunyuan-Large model was synthetic.

IN-CONTEXT LEARNING (ICL)
The ability of a model to learn or improve on tasks using the instructions and examples provided in the prompt, without requiring any further gradient updates. Sufficiently large models can perform this type of dynamic learning within their activations. ICL is essentially a learning algorithm inside of the learning algorithm—meta-learning. The simplest version of this, where a user gives the model examples of the task, is called few-shot prompting.

DWARKESH PATEL
Suppose this is the way things go, and we get these unhobblings…

LEOPOLD ASCHENBRENNER
And scaling. Scaling provides this baseline enormous force of improvement. GPT-2 was amazing [for its time]. It could string together plausible sentences, but it could barely do anything. It was kind of like a preschooler. GPT-4, on the other hand, could write code and do hard math, like a smart high schooler. This big jump in capability is explored in my essay series.[31] I count the orders of magnitude of compute and algorithmic progress.

Scaling alone, by 2027 or 2028, is going to do another preschool-to-high-school-sized jump on top of GPT-4. At a per-token level, the models will be incredibly smart. They'll gain more reliability. With unhobblings, they'll look less like chatbots and more like agents or drop-in remote workers.• That's when things really get going.

VIII.   DWARKESH PATEL

How do you make sense of the fact that when you give LLMs a lot of data in any specific domain, they tend to get better in just that domain? Wouldn't we expect a general improvement across all of the different areas?

DEMIS HASSABIS
Cofounder and CEO of Google DeepMind

You do sometimes get surprising improvement in other domains. For example, when these large models improve at coding, that can actually improve their general reasoning.[32] There is evidence of some transfer, although we would like a lot more evidence of that. But that's how the human brain learns, too. If we experience and practice a lot of things, like chess, creative writing, et cetera, we also tend to specialize and get better at that specific thing, even though we're using general learning techniques and systems in order to get better in that domain.••

TRANSFER
The ability to apply acquired knowledge effectively in different contexts, particularly to solve real-world problems beyond the original learning environment.

DWARKESH PATEL

As somebody who's been in this field for a long time and seen different trends come and go, what do you think the strong version of the scaling hypothesis gets right? What does it get wrong?

STRONG SCALING HYPOTHESIS
A current prevailing hypothesis in AI that holds that LLMs can achieve human-level intelligence with sufficient data and compute, with costs potentially in the range of trillions of dollars.

DEMIS HASSABIS

This is an empirical question right now. It was pretty surprising to almost everyone, including the people who first worked on the hypothesis, how far we've gotten. The models clearly have some form of concepts and abstraction. Five years ago, I would have said that we needed an additional algorithmic breakthrough to get that—maybe one more like how the brain works. I think that's still true if we want *explicit* abstract concepts, neat concepts, but it seems that these systems can already implicitly learn them.

We've got to push scaling as hard as we can. It's an empirical question whether we will hit a brick wall. No one knows. In the meantime, we should also double down on innovation. You can think of half of our effort as having to do with scaling. The other half has to do with inventing the next architectures and algorithms that will be needed,

ARCHITECTURE
The structure of a model, including how its components connect to one another and how it is trained. As of this writing, leading model architectures are still designed by humans.

•   That is, a substitute for a human performing a laptop job remotely from home.

••   Psychologists call improving in general by training in specific *far transfer*. It's the holy grail of education, in the sense that it is elusive. Niplav, "Transfer Learning in Humans."

SYMBOL GROUNDING
PROBLEM
A fundamental require-
ment and challenge for any
general AI system: the ability to translate between
sensory data and abstract
representations (for example,
between a set of written or
spoken instructions and the
corresponding objects and
sequences of actions in the
real world). To be effective,
the system must ground
symbols in the appropriate
real-world objects or
events. Hassabis is, in my
view, correct that this problem has more or less been
sidestepped. For example,
OpenAI's 2021 CLIP system
learned how to translate
between images and text
descriptions.

LABEL
In supervised learning,
a label is the correct or
desired answer, which is
applied to each input in the
training data. It is used to
guide the model's learning
process. Here, I'm referring
to the notion that at some
point humans won't be able
to produce useful labels
because we won't be able
to understand the outputs
of superhuman models.

knowing that larger and larger scaled models are coming down the line. My bet is that you need both.

I also think it's interesting and unexpected that these systems have some sort of grounding, even though they don't experience the world multimodally. I think we get some grounding through the RLHF feedback systems, because the human raters are, by definition, grounded, so their feedback is grounded too. Also, maybe language contains more grounding than we previously thought.

DWARKESH PATEL

Two things might change that would make grounding more difficult. One is that as these models get smarter, they're going to be able to operate in domains where we just can't generate enough human labels because we're not smart enough. If the model makes a million-line pull request, for example, how do we tell it whether this is within the con- straints of our morality and the end goal we wanted or not?

The other thing has to do with what you were saying about compute. So far, we've been doing next-token pre- diction, and in some sense, that's a guardrail. You have to talk as a human would talk, and maybe think as a human would think. Now, additional compute might be spent on reinforcement learning, which just *somehow* gets to the objective. We can't really trace how it got there. When you combine those two, how worried are you that the ground- ing goes away?

DEMIS HASSABIS

You have to have some grounding for a system to achieve goals in the real world. But these systems are becoming more multimodal [as of February 2024], ingesting things like video and audiovisual data as well as text data. The system correlates those things together. That is a form of proper grounding. So I do think our systems are going to start to understand the physics of the real world better.

IX.   TRENTON BRICKEN

Machine learning research is just so empirical. This is honestly one reason why I think our solutions might end up looking more brain-like than otherwise. Even though we wouldn't want to admit it, the whole community is doing a kind of greedy evolutionary optimization over the landscape of possible AI architectures. It's no better than evolution.

X.   DWARKESH PATEL

You're one of the only people outside of OpenAI who noticed the way AI was progressing in 2020, and you're maybe the only one who had a detailed model of scaling. What sort of process made you able to develop this model of what was happening with LLMs in your "Scaling Hypothesis" post? •

GWERN BRANWEN
Freelance writer and researcher

I was just a patient reader of everything, noting anomalies and then going back once in a while and checking again.

If I had to give an intellectual history, it would start in the mid-2000s, when I was reading [computer scientists Hans] Moravec and [Ray] Kurzweil. They were making the fundamental connectionist argument that getting enough computing power will result in a neural network that matches the human brain, and that until that computing power is available, trying to build AI is basically futile.

I found this argument very unlikely. It's very much a "build it and they will come" view of progress, which I did not think was correct. I thought it was ludicrous to suggest that simply because there's some supercomputer that matches the human brain in compute, that would summon the correct algorithm out of nonexistence. I thought, "You can't just buy a bunch of computers and expect to get an AI out. That's magical thinking." So, because I was super skeptical of the argument, I didn't pay too much attention to it.

But as part of my interest in transhumanism and AI risk, I was paying close attention to Shane Legg's blog posts, where he extrapolates the connectionist argument out with updated numbers, giving very precise predictions, like, "We're going to get the first general assistant around 2019, and then around 2025 we'll get agents and generalist capabilities, and by 2030 we should have AGI." I was, again, very skeptical. But along the way, [the semantic network] DanNet and [the image classification model] AlexNet came out—a very impressive success story of connectionism.[33] I thought, "Is it an isolated success story, or is it what Kurzweil and Moravec and Legg were predicting?" I started thinking that maybe scaling was not quite as stupid as I'd first thought.

It was this gradual trickle of drops hitting me as I went along. The dataset sizes kept getting bigger. The models kept getting bigger. The training runs crept up from using

CONNECTIONISM
A school of thought in cognitive science and AI that seeks to explain cognition in terms of neural networks. By the 2000s, the decades-long philosophical debate between the connectionists and the symbolists had largely subsided, just in time for deep learning to vindicate the connectionist argument. See, for example, Hans Moravec's "When Will Computer Hardware Match the Human Brain?" and Ray Kurzweil's *The Age of Spiritual Machines*.

•   Excerpted in the Appendix.

one cheap consumer GPU to two, and then to training on eight. The system's abilities kept getting broader and broader. Every few weeks, every few months, another drop. Finally, I went, "Maybe intelligence really is just a lot of compute applied to a lot of data. Huh. If that was true, it would have a lot of implications."

So there was no real eureka moment. I was just continually watching this trend that no one else seemed to see, except a handful of people like Ilya Sutskever or [computer scientist] Jürgen Schmidhuber. I just paid attention, noticing that the world looked more like [the connectionists'] world than like my world, where algorithms are super important and you need deep insight.

Then GPT-1 comes out. I was like, "Wow, this unsupervised sentiment neuron is learning on its own. That's pretty amazing." And then GPT-2 came out, and I was like, "Holy shit!" I looked at the prompting and the summarization, like, "Do we live in their world? *Can* we just keep scaling Transformers?"

Then GPT-3 comes down—the crucial test. Going from GPT-2 to GPT-3 is one of the biggest scale-ups in all of neural network history. If scaling was bogus, then the GPT-3 paper would be super unimpressive. Whereas if scaling was true, you would automatically get much more impressive results than GPT-2, guaranteed. I opened up the second page [of the paper] and I saw the few-shot learning chart, and I'm like, "Holy shit, we live in the scaling world. Legg and Moravec and Kurzweil were right!"

And then I turned to Twitter, and everyone else was like, "This shows scaling doesn't work! Why is GPT-3 not state of the art at everything?"[34] I was so angry at them that I had to write all this up.[35]

DWARKESH PATEL

In 2020, AI was already a thing. People were writing bestselling books about it. But none of those books were about scaling. What were people failing to account for?

GWERN BRANWEN

For the most part, they were suffering from two issues. First, they had not paid attention to all of the scaling results before 2020. They had not appreciated the fact that AlphaZero was discovered in part by doing Bayesian optimization on the hyperparameters and noticing that you can get rid of more and more of the tree search and get a better model. That was a critical insight that could only have been gained by having so much compute that you could train many,

TRANSFORMER
A modern neural network architecture notable for its parallel design and ability to learn context and relationships using a mechanism called self-attention. This attention mechanism dynamically assigns varying importance to different parts of the input data.

ALPHAZERO
Another game-playing AI developed by DeepMind that superseded AlphaGo. Unlike its predecessor, its training was pure self-play, using no human data. The system was also able to learn multiple games.

BAYESIAN OPTIMIZATION
A method of searching a space that is expensive to sample, such as the space of possible hyperparameter settings for a training run. Roughly, this involves creating a second model to predict how good a given setting will be, using this proxy to decide which settings to try next, and updating the model based on how good the setting actually was.

HYPERPARAMETER
A parameter that governs how a model is trained or operates. It's "hyper" because it governs the parameters (weights) of the model.

MONTE CARLO
TREE SEARCH
A method for identifying an appropriate sequence of actions by searching over an abstract decision tree. As an example, a game of chess can be represented as a branching tree of all possible sequences of moves. It is a powerful instance of symbolic AI, a rival of statistical machine learning that uses rules, logic, human representations, and explicit algorithms.

many versions and see the difference. And they simply did not know about the 2017 Baidu paper on scaling laws.[36] It should have been the paper of the year, but it didn't have any immediate impact. People were too busy discussing Transformers or AlphaZero.

Another issue is that they made the basic error I had made, thinking that algorithms are more important than compute. That's partly due to a systematic falsification of the actual origins of ideas in the research literature. Papers do not tell you where ideas come from; they just tell you a nice-sounding story about how something was discovered.[37] So even if you appreciate the role of trial and error and compute in your own experiments, you probably think, "I got lucky. Over in the next lab, they do things with the power of thought and deep insight." But it turns out that everywhere you go, compute, trial and error, and serendipity play enormous roles in how things actually happen.

Once you understand that, you understand why compute comes first. You can't do trial and error or serendipity at scale without it. You can write down all these beautiful ideas but you can't test them, or you can only test a few instances of it, so you typically find that it doesn't work, and you give up and do something else. Reading the old deep learning literature, you see all sorts of ideas that were completely correct but that no one could prove, like ResNets being first published way back in 1988 instead of 2015. The researchers didn't have the compute to train a version that would have worked.

Why believe that scaling was not going to work? Because you didn't notice the results that were key in retrospect. Another was BigGAN scaling to 300 million images.[38] There are still people today who will tell you with a straight face that GANs cannot scale past millions of images. If you don't know [otherwise], you could easily think, "GANs are broken. [We need a better algorithm.]" But if you do know that, then you think to yourself, "How can algorithms be so important when all these different generative architectures work, as long as you have lots and lots of GPUs?"

That's the common ingredient: lots and lots of GPUs. That's probably the root cause of not seeing scaling as a coherent paradigm and always [underrating it]. Even in 2020, you would still have AI people saying, "We'll get AGI in 2050." You could still think, very reasonably, that we still need lots and lots more incredible algorithmic breakthroughs [before we get AGI].

RESNET
Residual neural network. An important precursor to the Transformer architecture that uses residual connections between units (also known as skip connections) to let information flow between nonconsecutive layers.

XI. DWARKESH PATEL

Regarding your original point about LLMs needing episodic memory, you mentioned that these are problems that we can solve, not fundamental impediments. When you say that, do you think they will be solved through scaling, or do each of these require a specific fine-grained architectural solution?

SHANE LEGG

I think it'll be architectural, because current architectures don't really have what you need. They basically have a context window, which is very fluid, and they have weights, which [knowledge] gets baked into very slowly. To my mind, the model's activations are like working memory in your brain, and the weights are like the synapses in your cortex.

ACTIVATION
The value a model produces when processing a specific query, which depends on the weights it has learned during training and the inputs provided by the user; what gets input into the next layer of neurons in the model. Metaphorically, activations are like the electrical and neurotransmitter activity in the brain, or the model's active thoughts, associations, and goals.

Now, the brain separates these things out. It has a separate mechanism for rapidly learning specific information. That's a different type of optimization problem compared to slowly learning deep generalities.[39] There's a tension between the two. But you want both. You want to be able to hear someone's name and remember it the next day. You also want to be able to integrate information over a lifetime to see deeper patterns in the world.

These are quite different optimization targets, different processes, but a comprehensive system should be able to do both. So it's conceivable that you could build one system that does both. You can also see that because they're quite different things, it makes sense for them to be done differently. I think that's why the brain does it separately.

XII. DWARKESH PATEL

A big open question is whether reinforcement learning will allow these models to use self-play or synthetic data to get over data bottlenecks. It sounds like you're optimistic about this.

DEMIS HASSABIS

I'm very optimistic. First of all, there's still a lot more data that can be used, especially if one considers multimodal data. Obviously, society is adding more data to the internet all the time. There's a lot of scope for creating synthetic data. We're looking at that in different ways, partly through simulation—for example, using very realistic game environments to generate realistic data—but also through self-play. That's where systems interact with each other or converse with each other. It worked very well for us with AlphaGo and AlphaZero. We got the systems to play against each other, learn from each other's mistakes, and build up a knowledge base that way. There are some good analogies for that. It's a little bit more complicated to build a general kind of world data.

DWARKESH PATEL

How do you get to the point where the synthetic data the models are outputting on self-play is not just more of what's already in their dataset but something they haven't seen before? Something that would actually improve their abilities.

DEMIS HASSABIS

There's a whole new science needed there. This is important for things like fairness and trying to remove bias from the system and making sure that the dataset is representative of the distribution you're trying to learn. We're still in the nascent stage of [optimal] data curation and data analysis and analyzing the holes in our data distribution. There are many tricks one can use, like overweighting or replaying certain parts of the data. Or, if you identify some gap in your dataset, that's where you put synthetic generation to work.

XIII.   DWARKESH PATEL

What should we make of the fact that these models require so much training and the entire corpus of internet data in order to become merely *sub*human? Should we be worried about how inefficient these models seem to be?

DARIO AMODEI

That's one of the remaining mysteries. One way you could phrase it is that the models are maybe two to three orders of magnitude [100x to 1,000x] smaller than the human brain, while at the same time being trained on three to four orders of magnitude [1,000x to 10,000x] more data. Compare the number of words a human sees as it's developing until age 18. I think it's in the hundreds of millions.[40] Whereas for the models, we're talking about trillions.

What explains this? The models are smaller than brains, so they need a lot more data.• Or perhaps the analogy to the brain is not quite right or is breaking down. There's some missing factor. This is just like in physics, when we

---

•   A lesser-known scaling law is that larger models are actually more sample efficient—they learn more from each data point than smaller models. See Kaplan et al., "Scaling Laws."

couldn't explain the Michelson-Morley experiment• or one of the other 19th-century physics paradoxes. It's something we don't quite understand. Humans see so little data and they still do fine.

One theory could be that it's our other modalities that do it. How do we get 10^{14} bits into the human brain? Maybe most of it is images.•• Maybe a lot of what's going on inside the human brain is that our mental workspace involves simulated images, or something like that.

Honestly, we have to admit it's weird. It doesn't match up. This is one reason I'm a bit skeptical of biological analogies. I thought in those terms five or six years ago. Now that we have these models in front of us, it feels like the evidence from these analogies has been screened off by what we've actually seen. What we've seen are models that are much smaller than the human brain, and yet they can do a lot of the things that humans can do. And yet, paradoxically, they require a lot more data to do those things.

Maybe we'll discover something that makes it all efficient. Maybe we'll understand why the discrepancy is present. At the end of the day, I don't think it matters if we keep scaling the way we are. What's more relevant at this point is just measuring the abilities of the model and seeing how far they are from humans' abilities. They don't seem terribly far to me.

•   A crucial 1887 experiment that disproved the aether theory of light propagation, paving the way for the discovery of special relativity.

••   Here's a rough comparison: The eyes receive around 10 million bits of information per second. An ordinary person can read five words per second, which, in English, is about 50 bits per second—thousands of times slower. Spoken language tends to be about 39 bits per second. The skin is another high-bandwidth channel, processing perhaps 1 million bits per second, though it still doesn't approach the optic information rate. One caveat is that most optic information doesn't reach the brain. Markowsky, "Information Theory"; Coupé et al., "Different Languages, Similar Encoding Efficiency"; Koch et al., "How Much the Eye Tells the Brain."

Chapter 2
Evals

What can LLMs do?
How do we know?

GPT-4 can solve novel and difficult tasks that span mathematics, coding, vision, medicine, law, psychology, and more, without needing any special prompting... We believe that it could reasonably be viewed as... showing sparks of artificial general intelligence.
—Sebastien Bubeck et al., 2023

The bigger models? They smooth out all the trivial failings. They unwrinkle the creases. They babble on for longer and longer stretches without falling flat on their face. But eventually they fall, and they fall just as *hard*.
—Nostalgebraist, 2021

What has scaling bought us? What, exactly, can current models do, and how do researchers work that out? These questions fall under "evals": the science of evaluating AI capabilities.

Figure 2 in the previous chapter showed how models' loss improves with scaling. But the pretraining loss is not what we care about. We want the models to complete real tasks, like writing code, doing our grad school homework, or, eventually, making money for us while we kick back.[41] A benchmark is intended to show that the model generalizes—that it goes beyond memorizing the data it is shown and finds an algorithm that reflects the true structure of the task.

Figure 4 shows results from one core benchmark for such general ability: BIG-Bench, a collection of 200 tasks selected to be hard for language models to accomplish. The models' performance on these tasks, which include identifying anachronisms, reading ASCII art, and determining the author of a passage, also improves smoothly with vast increases in training compute.•

The goal of this chapter is to explain the significance of beating a benchmark like BIG-Bench—and why, despite years of solid improvement, we're still often surprised by what new models can do.••

The space of possible inputs for a modern model is impossible to check exhaustively, so evals are only ever

PRETRAINING LOSS
A measure of how well a model predicts the next token on average during the initial unsupervised training phase.

BENCHMARK
The standard method for evaluating a model's capabilities and performance. Benchmarks typically consist of tasks that are challenging for models to perform, often in the form of fixed datasets.

• However, these tasks are contrived and short-horizon (meaning they require few sequential steps), which may disqualify them from being considered real-world tasks.

•• One challenge is that benchmarks can't be taken at face value due to the intense academic and commercial incentives to subvert them, and because there are many ways to fool oneself or cheat. However, they still provide some evidence of how model performance is improving. Leech et al., "Questionable Practices in Machine Learning."

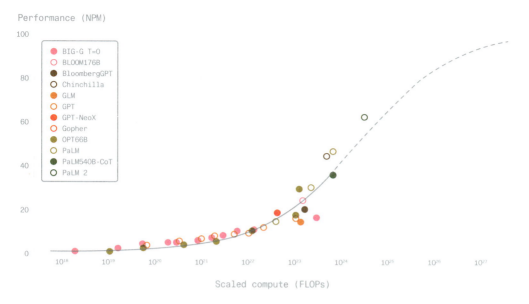

Performance (NPM)

Scaled compute (FLOPs)

Figure 4. Scaling of downstream task performance on BIG-Bench, which consists of over 200 tasks, in 2022. The dotted line represents Epoch's projection of future BIG-Bench performance given larger training runs; it turns out that their model actually underestimated the performance of PaLM 2. Owen, "Performance in Language Modeling Benchmarks."

sampling incompletely from a very large distribution.•
Anthropic cofounder Jack Clark calls this uncertainty—
the fact that we often only work out how strong a model is
after its release—the capability overhang. He asks, "What
about all the capabilities we don't know about because we
haven't thought to test for them?"••

•   Consider that models now allow for inputs of 128,000 tokens
        or more—the size of an entire book—and that each of these
        tokens can take any of 100,000 values. This illustrates how the
        number of possible prompts increases exponentially. van Gilst,
        "Analyzing GPT-4 Tokens."

••   Vincent, "AI Is Finally Mainstream." A simple example of
        the capability overhang is MedPrompt, a heavily engineered
        approach to sampling GPT-4 to get it to answer medical
        questions. Researchers found that simply changing how the
        questions were worded improved the answer accuracy of a
        larger, unspecialized model by 4 percent—enough to outper-
        form all specialized medical AI systems. More generally, models
        can learn in context (that is, they can use information provided
        by a user to rapidly improve on many tasks). This illustrates
        how improving an LLM does not always require scaling it.
        Other methods include giving it access to scaffolds and agentic
        workflows. Nori et al., "Can Generalist Foundation Models
        Outcompete Special-Purpose Tuning?"; Ng, "Issue 242."

One might think that LLMs would be bounded by human performance, since they are trained on human data and haven't seen any superhuman performance to imitate. But in some domains, like chess, we do have examples of superhuman performance to train on. And in some cases, models go beyond their training data.• As mentioned in the Preface, in terms of speed, breadth of knowledge, and accuracy at predicting the next word in a sentence, they are already superhuman.[42]

We always have two options when faced with the implications of a test result. If a test finds a system to be intelligent, we can reject the test ("Clearly, the test is too easy," or "The model just pattern matched"). If a system is found to exhibit human-level intelligence, we can either raise our estimation of the system's intelligence or lower our estimation of human intelligence.[43]

Evals play a bigger role than simply estimating performance; they also help us work out what *kind of thing* LLMs

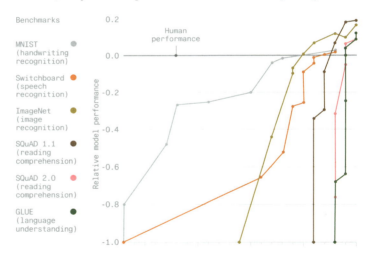

Figure 5. The time between a benchmark being created and being solved (that is, when the model is able to match human performance on ideally unseen test data) for six notable benchmarks. We see that these benchmarks saturate, or reach maximum performance, far more quickly in recent years.

• Most existing examples of a model generalizing from training data are for closed-world games. A 2024 paper demonstrated this effect for chess, while the Atari-playing Decision Transformer also occasionally transcended its training data. Still, this is a significant finding—it shows that models trained on the traces of explicit search can reason without relying on explicit search algorithms. Ruoss et al., "Grandmaster-Level Chess without Search"; Zhang et al., "Transcendence"; Chen and Lili, "Decision Transformer."

are. We first develop a theory of what these models are doing ("They're doing approximate retrieval of the internet"; or "They're doing linearized subgraph matching"; or "They're searching over a continuous space of implicit Turing machines"), then we use evals to test whether they fail or excel at the tasks implied by the theory.[44]

In this chapter, we examine how models behave, asking what they can and can't do and how we can measure their progress. We also discuss how we should define intelligence in the first place, so that we know when we've found it.

1.   DWARKESH PATEL
What will the ultimate grand parsimonious theory of intelligence look like?

GWERN BRANWEN
Freelance writer and researcher
The 10,000-foot view that the success of scaling points to is that all intelligence is, is search over Turing machines. Recall that anything that can happen can be described by Turing machines of various lengths.[45] When we scale a model, all we're doing is searching over more and longer Turing machines and applying them [to each task or thought]. Otherwise, there is no general master algorithm. There is no magic transfer to all tasks. There is no intelligence fluid. It's just a tremendous number of special cases that we learn and encode in our brains.

DWARKESH PATEL
When I think about the ways humans differ in intelligence, it seems much more like a difference in general horsepower, whereas the difference between models is just that bigger models have a larger representation of ever-sparser features and a longer tail of things that might come up. It's not just that my smarter friends know more things; it's like they've got more juice. Is that represented by this Turing machine picture?

GWERN BRANWEN
Sure. When we talk about people having more intelligence, it's just that they have more compute to do longer searches over larger or more Turing machines. Nothing more than

* This is the conclusion of Marcus Hutter's work on AIXI, a mathematical theory of universal intelligence. Hutter, "Universal Artificial Intelligence."

that. You can always extract a small solution to a specific problem from any learned model or brain, because all the large brain is doing with the extra compute is finding it.

That's why you never find an IQ gland. There is no region in the brain where, if you remove it, you eliminate fluid intelligence. That turns out not to exist, because what your brain is doing is learning specialized solutions to problems and then recombining them into fluid intelligence.[46] That's intelligence. That's all the large model is: a gigantic ensemble of small models tailored to the ever-escalating number of problems you've been feeding them.

II.   DWARKESH PATEL

Let's talk about the idea that LLMs are fitting some manifold underlying the input data. A reductionist way to talk about the human brain is that it's just axons firing at each other. But we don't care about the reductionist explanation. We care about what happens at the macroscopic level when these things combine.• The GSM8K benchmark is grade school math—problems that a smart high schooler would be able to solve. Models get 95 percent on it. They nail it.

GSM8K
A benchmark of 8,500 grade school math word problems used by LLM developers to evaluate and compare models' problem-solving and reasoning abilities.

FRANÇOIS CHOLLET
Creator of ARC-AGI and Keras

Sure, that's a memorization benchmark.

DWARKESH PATEL

Here's a question from GSM8K: "Thirty students are in a class. One-fifth of them are 12-year-olds. One-third are 13-year-olds. One-tenth are 11-year-olds. How many of them are not 11, 12, or 13 years old?"

•   Similarly, saying that models are "just a pile of linear algebra" or "just compressing the training data" is moot—it's the wrong level of analysis. A classic taxonomy from neuroscience, Marr's levels of analysis, clarifies this. A brain, computer, or LLM can be analyzed on at least three levels:

1.   Objective: The computational problem the system is solving.
2.   Algorithm: The abstract plan the system uses to solve that problem.
3.   Implementation: How that algorithm is implemented on hardware.

Many severe mistakes arise from being unclear about these levels. Calling an LLM "just linear algebra" mistakes level 3 for level 2. Calling an LLM "just a compression of its training data" mistakes level 1 for level 2. You can detect a Marr violation quite reliably by its use of the word "just." Marr, *Vision*.

I agree this isn't rocket science. A smart high school kid should be able to solve it. But as for being "just memorization," the model still has to reason through how to think about fractions, the context of the whole problem, and then combine different calculations to write the final answer.

FRANÇOIS CHOLLET

It depends on how you want to define reasoning. There are two definitions you can use. One is, "I already have a set of program templates, the structure of the puzzle, which can generate its solution. I'm going to identify the right template, which is in my memory, input new values, run the program, and get the solution." You could say this is reasoning. Sure.

Here's another definition of reasoning: the ability to synthesize a new program on the fly based on bits and pieces of existing programs. That's dramatically harder to do than just fetching the right memorized program and reapplying it.

DWARKESH PATEL

Maybe we are overestimating the sample efficiency of humans at that second kind of reasoning. Humans also need training. We have to drill in pathways of reasoning through [seeing] certain kinds of problems.

Take math. It's not like you can show a baby the axioms of set theory and now they know math. As they're growing up, you have to teach them years of pre-algebra. Then a year of teaching them drills and going through the same kind of problem in algebra. Then geometry, precalculus, calculus. Isn't that the same kind of thing? You can't just see one example and now you have the program. You have to drill it. These models also had to drill it with a bunch of pretraining data.

FRANÇOIS CHOLLET

Sure. In order to do on-the-fly program synthesis, you need some building blocks to work from. Knowledge and memory are tremendously important in the process. I'm not saying it's memory versus reasoning. In order to do effective reasoning, you need memory.

DWARKESH PATEL

But then this sounds compatible with the idea that LLMs reason. Through seeing a lot of different kinds of examples, these things can learn to reason within the context of those examples. We can also see this within bigger and bigger models. I gave an example of a high-school-level math problem. Let's say a model smaller than GPT-3 couldn't do that at all. As these models get bigger, they seem to be able to pick up bigger and bigger patterns.

FRANÇOIS CHOLLET

It's not really a size issue. It's more like a training data issue in this case.

DWARKESH PATEL

Well, bigger models can pick up these kinds of circuits. Smaller models apparently don't do a good job of that, even if you were to train them on this kind of data. Doesn't that suggest that as you have bigger and bigger models, they can pick up bigger and bigger pathways or more general ways of reasoning?

FRANÇOIS CHOLLET

Absolutely.

DWARKESH PATEL

But then isn't that intelligence?

FRANÇOIS CHOLLET

No, it's not. If you scale up a database and keep adding more knowledge and program templates to it, then, sure, it becomes more and more skillful. You can apply it to more and more tasks. But general intelligence is not task-specific skill scaled up. General intelligence is the ability to approach any problem, any skill, and very quickly master it using very little data. This is what makes you able to face anything you might ever encounter. Generality is not specificity scaled up. It is the ability to apply your mind to anything at all, to arbitrary things. This fundamentally requires the ability to adapt, to learn on the fly efficiently.

DWARKESH PATEL

My claim is that by doing pretraining on bigger and bigger models, you are gaining that capacity to generalize very efficiently.

Let me give you an example. Your own company, Google, in their paper on Gemini 1.5, had this very interesting example.[47] They would give the model, in context, the grammar book and the dictionary of a language that has fewer than 200 living speakers. It's not in the pretraining data. They give it the dictionary, and then it's basically able to speak this language and translate to it.

If you showed me an English–Spanish dictionary, I'm not going to be able to pick up how to structure sentences and say things in Spanish. Because of the representations the model has gained through this pretraining, it is able to learn a new language extremely efficiently. Doesn't that show that this kind of pretraining does increase your ability to learn new tasks?

FRANÇOIS CHOLLET

If you were right, LLMs would do really well on ARC puzzles, because ARC puzzles are not complex.[48] Each of them requires very little knowledge. Each of them is very low complexity. They're extremely obvious for humans.• Even children can do them. But LLMs cannot. Even LLMs that have 100,000 times more knowledge than you *still* cannot. The only thing that makes ARC special is that it was designed with this intent to resist memorization. This is the huge blocker for LLM performance.

If you look at LLMs closely, it's pretty obvious that they're not really synthesizing new programs on the fly to solve the task that they're faced with. Instead, they're reapplying things that they've stored in memory. For instance, LLMs can solve a Caesar cipher, transposing letters to code a message. That's a very complex algorithm, but it comes up quite a bit on the internet. They've basically memorized it. What's really interesting is that they can do it for a transposition length of three or five, because those are very common numbers in examples provided on the internet. If you try to do it with an arbitrary number like nine, it's going to fail.•• It does not encode the generalized form of the algorithm but only specific cases. If it could actually synthesize the solver algorithm on the fly, then the value of n would not matter at all because it does not increase the problem complexity.

CAESAR CIPHER
A simple encryption method in which the letters of the plaintext are shifted by a fixed amount, for example C → F, A → D, E → H, and so on.

DWARKESH PATEL

I think this is true of humans as well.

• Some humans, anyway. Sören Mindermann (@sorenmind), "Francois Chollet says the ARC challenge for AI is easy even for 5 year olds. Can a 5 year old help me with this please?" X, June 19, 2024, https://x.com/sorenmind/status/1803558237909782795.

•• As of 2024, this is no longer true. (A complicating factor is that recent LLM systems, such as GPT-4, include tools like Code Interpreter and Advanced Data Analysis, which the LLM can use for challenging tasks.) ChatGPT, "Caesar Cipher."

FRANÇOIS CHOLLET

Humans use memorization and pattern matching all the time, of course, but humans are not limited to pattern matching. We have this very unique ability to adapt to new situations on the fly. This is what enables you to navigate every new day in your life.

DWARKESH PATEL

How do you explain how Gemini 1.5 was able to learn a language in context, including its complex grammar? Doesn't that show that models can pick up new knowledge?

FRANÇOIS CHOLLET

I would assume that it has simply mined it from its unimaginably vast training data. It has mined the required template and is just reusing it. We know that LLMs have poor ability to synthesize new program templates like this on the fly, or even adapt existing ones. They're very much limited to fetching [data].

III.   DWARKESH PATEL

If you'd told me in 2018 that in 2023 we'd have models like Claude 2 that can write theorems in the style of Shakespeare, explain whatever theory you want, ace standardized tests with open-ended questions, and do all kinds of really impressive things, I would have said, "You have AGI. You clearly have something that is of human-level intelligence."

While impressive, it seems we're not at human level yet, at least in the current generation, and potentially for generations to come. What explains this discrepancy between super-impressive performance in these benchmarks versus it lacking general intelligence?

DARIO AMODEI
CEO of Anthropic

I was surprised as well. When I first looked at GPT-3, my general sense was, it seems like [the models have] really grasped the essence of language. I'm not sure how much we need to scale them up. Sure, we can scale pretraining a bunch more, but I wonder if it's more efficient to start adding

---

• Pretraining an LLM now involves approximately 10 trillion gradient update steps. In contrast, RLHF is likely closer to 100 million steps because it requires some manually labeled data, and is therefore expensive. It's also plausible that the success of OpenAI's o1 on reasoning tasks can be attributed to increasing RL's share of the total training effort. Rafailov et al., "Direct Preference Optimization"; Dubey et al. "Llama 3."

on other objectives, like RL. Maybe doing as much RL as you've done pretraining on a 2020-style model is the way to go.● Scaling pretraining up will keep working. But is that really the best path? I don't know. But it just keeps working.

If you look back to these diagrams that are like, "Here's the village idiot, here's Einstein, here's the scale of intelligence," and the village idiot and Einstein are very close to each other [see Figure 6]—maybe that's still true in some abstract sense, but it's not really what we're seeing.[49] We're seeing that the human range is pretty broad, and [the models] don't hit the human range in the same place or at the same time for different tasks.

Figure 6. One possible way to graph the relative intelligence of various beings. The implicit claim is that intelligence can be projected onto a single dimension and that there may be beings far more intelligent than the smartest humans, making the difference between the most and least intelligent humans seem small. Yudkowsky, "My Childhood Role Model." ●●

[You can tell the model] to write a sonnet in the style of Cormac McCarthy. I'm not very creative, so I couldn't do that, but that's a pretty high-level human skill. The model is starting to get good at constrained writing: "Write a page about topic X without using the letter E." The models

●●   In personal correspondence in late 2024, Yudkowsky said, "How long it's taking to get from GPT-4 to Einstein is evidence against the distance being as narrow as I drew it then. We are learning something about how there's more distance inside humanity than my younger, more idealistic self believed and, I'd now say, merely hoped. I don't think we're learning anything startling about diversity or compartmentalization of cognitive skills; my younger self already expected and predicted that AIs would have a very inhuman balance of abilities. That the leading AI tech for many tasks (for example, coding) is AI trained on generalized tasks and then asked to do that particular thing (rather than an AI trained on just coding) validates the underlying notion of 'general intelligence' as it applies to an interspecies scale. It's hard to have an AI that's good at only one task if the task is sufficiently deep, because the sufficiently deep tasks require deep capabilities that generalize."

might be superhuman or close to superhuman at that. But when it comes to proving relatively simple mathematical theorems, they're just starting to do the basics. They make really dumb mistakes sometimes, and they lack any kind of broad error correction or ability to do an extended task.

So it turns out that intelligence isn't a single spectrum. There are a bunch of different areas of domain expertise. There are a bunch of different kinds of skills. Memory is different; it's not complicated. But to the extent that memory is even on the spectrum, the spectrum is also wide. If you had asked me 10 years ago, that's not what I would have expected, but that's the way it's turned out.

IV.   DWARKESH PATEL

MASSIVE MULTITASK
LANGUAGE
UNDERSTANDING (MMLU)
A popular benchmark for
evaluating LLM capabilities,
comprising 16,000 multiple-
choice questions across
57 academic disciplines,
including math, philosophy,
medicine, and law.

We measure the performance of these large language models on MMLU and other benchmarks. What is missing from the benchmarks we use currently? What aspect of human cognition do they not measure adequately?

SHANE LEGG
Cofounder and chief AGI scientist at Google DeepMind

They don't measure things like understanding streaming video. These are *language* models. They don't have things like episodic memory. Humans have a working memory, for things that have happened quite recently, and then we have a cortical memory, things being stored in our cortex. But there's also a system in between: episodic memory, in the hippocampus. It's for learning specific things very rapidly. So if you remember some of the things I say to you tomorrow, that'll be your episodic memory. Our models don't really have that kind of thing, so we don't really test for it. We just try to make the context windows, which is more like working memory, longer and longer to compensate.

It's a difficult question because the generality of human intelligence is very broad. You have to go into the weeds of trying to find out if there are specific types of things that are missing from existing benchmarks or different categories of benchmarks that don't currently exist.

DWARKESH PATEL
Would it be fair to call episodic memory the root of human sample efficiency, or is that a different thing?

SHANE LEGG
It's very much related to sample efficiency. It's one of the things that enables humans to be very sample efficient. Large language models have a certain kind of sample

efficiency, because when something is in their context window, that biases the distribution to behave in a different way. That's a very rapid kind of learning. There are multiple kinds of learning, and the existing systems have some of them but not others. It's a little bit complicated.

DWARKESH PATEL

Is it a fatal flaw of deep learning models that it takes them trillions of tokens to learn, or is this something that will be solved over time?

SHANE LEGG

The models can learn things immediately when they're in the context window. Then they have this longer process, where you actually train the base model. That's when they're learning over trillions of tokens. What I'm getting at is that they're missing something in the middle.

I don't think it's a fundamental limitation. What's happened with large language models is that something fundamental has changed. We know how to build models that have some degree of understanding of what's going on. That did not exist in the past. We've got a scalable way to do this now, which unlocks lots and lots of new things. We can look at things that are missing, such as this episodic memory type thing, and we can start to imagine ways to address that.

My feeling is that there are relatively clear paths forward now to address most of the shortcomings we see in the existing models. Whether it's about delusions, factuality, the type of memory and learning they have, understanding video, all sorts of things like that, I don't see big walls in front of us. I just see that with more research and work, these things will improve and probably be adequately solved.

DWARKESH PATEL

Let's go back to the original question of how you measure when human-level AI has arrived, or when we've gone beyond it. As you mentioned, there are other benchmarks you can use, and other sorts of traits. Concretely, what would a model have to do for you to be like, "Okay, we've reached human level"? Would it have to beat *Minecraft* from start to finish? Would it have to get 100 percent on MMLU?

SHANE LEGG

There is no one test that would do it, because that's the nature of general intelligence. I'd have to make sure it could do lots of different things and didn't have a gap.

We already have systems that can do very impressive categories of things to human level or even beyond. I would want a whole suite of tests that I felt was very comprehensive. Furthermore, when people come in and say, "Okay, it's passing a big suite of tests. Let's take an adversarial approach to this. Let's deliberately try to find examples [of tasks] where people can typically do this, but the machine fails..." When those people cannot succeed, I'll go, "Okay, we're probably there."

V.   DWARKESH PATEL
Google DeepMind is at the frontier of having agents that can think through different steps to get to an outcome, and has been for many years with systems like AlphaZero. Is there a path for LLMs to have tree search on top of them?

DEMIS HASSABIS
Cofounder and CEO of Google DeepMind
I think that's a super-promising direction. We've got to carry on improving the large models. We've got to carry on making them more and more accurate predictors of the world, making them more and more reliable world models. That's clearly a necessary, but probably insufficient, component of an AGI system.

On top of that, we're working on things like AlphaZero-like planning mechanisms that make use of that model in order to make concrete plans to achieve certain goals in the world. Perhaps chaining thought, lines of reasoning, together and using search to explore massive spaces of possibility is missing from our current large models.

DWARKESH PATEL
How do you get past the immense amount of compute these approaches tend to require? Even AlphaGo was pretty expensive, because you had to run [a search] on each node of the tree. How do you anticipate that will be made more efficient?

DEMIS HASSABIS
Moore's law tends to help. Every year, more computation comes in. But we focus a lot on sample-efficient methods and reusing existing data, things like experience replay.

Also, the better your world model is, the more efficient your search can be. One example I always give is Alpha-Zero, our system to play Go and chess and any game. It's stronger than a human world champion in all these games, and it uses a lot less search than a brute-force method like

EXPERIENCE REPLAY
A technique to stabilize the training of RL policies by having the model store past episodes and learn from them multiple times.

DEEP BLUE
The first superhuman chess system, released in 1997. It used summary data from 700,000 grandmaster games and tree search to identify strong moves, performing around 200 million board evaluations per second.

STOCKFISH
An open-source chess engine that is currently recognized as one of the strongest in the world. After being defeated by DeepMind's AlphaZero, its developers switched to a hybrid of hand-crafted functions and neural networks, and later to a pure neural policy.

Deep Blue. A traditional Stockfish or Deep Blue system would look at millions of possible moves for every decision. AlphaZero and AlphaGo may look at tens of thousands of possible positions in order to make a decision about what to move next. A human grandmaster or world champion probably only looks at a few hundred moves in order to make their very good decision about what to play next.

That suggests that the brute-force systems don't have any real model other than the heuristics about the game. AlphaZero has quite a decent model, but the top human players have a much richer, much more accurate model. That allows them to make world-class decisions with a very small amount of search. So there's a trade-off there. If you improve the model's [understanding of the game], then your search can be more efficient, and therefore you can get further with your search.

VI.   DWARKESH PATEL

People have been testing whether these models are overfit to the benchmarks. Scale AI recently did this with GSM8K. [See Figure 8.] They basically replicated that benchmark

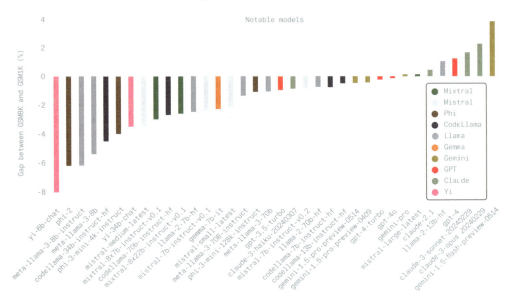

Figure 7. The LLM reasoning gap on the GSM8K benchmark (that is, how much the models' performance changed after ruling out any possibility of memorizing the answers). In 2024, researchers created a new test set for GSM8K and found that the performance of many LLMs decreased dramatically. However, Claude 3 actually improved its score, resulting in a negative reasoning gap, indicating better performance. Zhang et al., "Large Language Model Performance on Grade School Arithmetic."

OVERFITTING
A modeling error in which a
system memorizes the training set instead of learning
the underlying principles
that generated it—essentially,
learning from noise. Here, I'm
actually talking about training contamination, which is
when a model sees the test
answers during training and
thus "cheats" by already
knowing the answer key.

but with new questions.[50] Some models, like Mistral, were extremely overfit on the benchmark. Frontier models like Claude and GPT actually did as well on their novel benchmark as they did on the specific questions that were in the existing public benchmark.

FRANÇOIS CHOLLET

There's no question about whether current models are trained on the test data. They are, because they train on all of GitHub.

VII.   DWARKESH PATEL

Famously, in the social sciences, results are really hard to replicate. There's a question about how much of the science is real versus manufactured. When you look at the average ML paper, does it feel like a really solid piece of literature, or does it often feel p-hacked?

P-HACKING
The practice of manipulating data to make it appear
statistically significant. Here,
it stands in for questionable
research practices more
broadly.

JOHN SCHULMAN
Senior researcher at Anthropic

Everyone has their complaints about the ML literature. Overall, I think it's a relatively healthy field, especially compared to some others, like the social sciences. It's largely grounded in practicality, in getting things to work. It's also accepted that you don't just report someone's [results], you try to reimplement their method and compare it to your method on the same training dataset. If you publish methods that are really hard to implement or finicky, they'll tend to get forgotten.

There are also various unfavorable incentives. People are incentivized to make the baseline methods they're comparing to worse. There are other mild pathologies, like trying to make your methods seem mathematically sophisticated. But overall, the field makes progress. I would like to see a little bit more science and more attempts to understand things rather than just hill-climbing on benchmarks and trying to propose new methods. There's been a decent amount of that recently. We could use more.

VIII.   DWARKESH PATEL

Your claim is that AI agents haven't taken off because of their low reliability rather than their long-horizon task performance. But isn't lack of reliability exactly the difficulty with long-horizon tasks? You have to do 10 or 100 steps in a row, diminishing the overall task reliability. The whole thing gets multiplied together and the whole task becomes much less likely.

SHOLTO DOUGLAS
Reinforcement learning infrastructure lead at Anthropic

That is exactly the problem. But the key issue is that your base task solve rate is 90 percent. If it was 99 percent, it wouldn't be a problem.•

This also hasn't been properly studied. If you look at the academic evals, [they test models on single-step] problems. MATH, for example, is one typical math problem. You were beginning to see evals looking at this properly via more complex tasks like SWE-bench, where they take a whole bunch of GitHub issues. That is a reasonably long-horizon task, but it still [takes a skilled human less than an] hour, as opposed to a multihour or multiday task.

One thing that will be really important to do next is to understand better what the success rate over long-horizon tasks looks like. That's important to understand what the economic impact of these models might be and to properly judge increasing capabilities. You'd cut down the tasks and the inputs/outputs involved into minutes or hours or days and see how good it is at successively chaining and completing tasks at those different resolutions of time. That tells you how automated a job family or task family will be in a way that MMLU scores don't.

IX.   DWARKESH PATEL

What are you keeping track of while you're doing long-horizon RL? How would you notice a discontinuous jump in capabilities before you deployed the systems broadly?

JOHN SCHULMAN

You would want a lot of evals running during the training process.

DWARKESH PATEL

What, specifically? Does it make sense to train a long-horizon RL knowing that this is something that could happen, or is it very low probability?

JOHN SCHULMAN

You'd want to be pretty careful when you do this kind of

MATH
A popular benchmark used to test LLMs on challenging high school mathematics problems.

SWE-BENCH
A benchmark consisting of real software maintenance and feature request tasks.

LONG-HORIZON REINFORCEMENT LEARNING
An approach to post-training LLMs that uses reinforcement learning to create long chains of rewards and penalties, thereby instilling the ability to execute long chains of actions.

•   Consider a task that consists of 10 necessary steps. If an LLM succeeds at each step 99 percent of the time, its overall success rate will be approximately 90 percent (0.99¹⁰). However, if it plateaus at 90 percent success for each step, the overall task success rate drops to about 35 percent (0.90¹⁰), which is often practically unusable.

training if you saw potentially scary capabilities. It's not something we have to be scared of right now, because right now it's hard to get the models to do anything coherent.

If they started to get really good, we would want evals that test for misbehavior. We'd want to check that they're not going to turn against us. You might also want to look for discontinuous jumps in capabilities.

You'd also want to make sure that whatever you're training on doesn't have any reason to make the model turn against you. That doesn't seem like the hardest thing to do. The way we train them with RLHF feels very safe, even though the models are very smart. The model is just trying to produce a message pleasing to a human. It has no concern about anything else in the world other than whether the text it produces is approved.

Obviously, if you were doing something where the model has to carry out a long sequence of actions involving tools, then it might have some incentive to do wacky things that wouldn't make sense to a human in producing its final result. However, it wouldn't necessarily have an incentive to do anything other than produce a very high-quality output. Of course, if you assigned it a task like "make money," then maybe that would lead to some nefarious behavior as an instrumental goal.

Chapter 3
Internals

What's going on inside
these models?

One might be tempted to define thinking as consisting of "those mental processes that we don't understand." If this is right, then to make a thinking machine is to make one which does interesting things without our really understanding quite how it is done.
—Alan Turing, 1952

For what is the heart, but a spring; and the nerves, but so many strings; and the joints, but so many wheels.
—Thomas Hobbes, 1651

One view of intelligence is that it is simply the ability to solve a great variety of problems.• Under this pragmatic definition, LLMs would clearly count as intelligent. But even if we grant that LLMs demonstrate some form of intelligence, we're left with another question: In performing as well as they do, do they reason? Do they represent the structure of the world in a robust way? How, precisely, do they obtain their output?

We don't know. To understand why we don't know, we have to distinguish the model's architecture—the way a researcher sets up the blank model before training it—from its training objective—which, in pretraining, is to successfully predict the next token in a document—and its internal representations—the concepts, algorithms, features, and heuristics learned during training. Training encodes these representations in the final values of the weights, which dictate its outputs.

We understand an LLM's architecture and objective just fine, but these learned and massively distributed representations remain unclear. Does an LLM *reason* in some strict sense, or does it merely skillfully exploit masses of statistical associations? Relatedly, does it have a world model—a stable, self-consistent map of reality it manipulates to answer questions?[51] Opinions differ.

The skeptical hypothesis is that the models are memorizing answers or pattern-matching chains of words—essentially, babbling.•• But we know that they

DISTRIBUTED
Noncentralized; a concept that is spread across thousands of neurons instead of being represented by a single neuron. Each neuron in a model is thus polysemantic, involved in storing and manipulating multiple concepts.

• This view is common in AI. For example, David Hambrick, Alexander Burgoyne, and Erik Altmann write in the 2019 *Cambridge Handbook of Intelligence*, "The ability to solve problems is not just an aspect or feature of intelligence—it is the essence of intelligence."

•• In humans, verbal and mathematical intelligence correlate positively, though some people exhibit a strong imbalance favoring one over the other.

can reason.● We also know that they memorize training data, as revealed by the collapse in performance when tested on newly collected data.[52] More recent evaluations aim to test this reasoning gap, or the difference between the model's ability to solve new problems and its skill at simply recalling or generalizing from old answers.[53]

For obvious reasons, this also opens up a more philosophical line of inquiry. How is intelligence implemented in the brain? Can an LLM be said to think? Does it have a mind?[54] Interpretability is the science of working out what a model's weights and activations mean in human terms. In other words: How do the models think, *if* they think?

Recent work has made progress in this domain. For example, the following are abbreviated reconstructions of some of the concepts that activate most strongly inside an LLM (here, Claude 3.5 Sonnet) when it is asked questions about itself:

- "Referring to an android, robot, AI, or machine entity using gendered pronouns like 'she,' 'her,' or 'his.'"
- "Mentions of artificially created, programmed, or robotic entities like androids and cyborgs."
- "Concept of artificial intelligence becoming self-aware, transcending human control, and posing an existential threat to humanity."
- "Concepts related to entrapment, containment, or being trapped or confined within something like a bottle or frame."[55]

All this suggests that an LLM's internal representation of an LLM draws heavily on our own tropes and preconceptions about AIs.

This chapter is by far the most technical in the book. The exchanges delve into the details of model internals, including how language models could perform complex reasoning, how intelligence arises both in neural networks and in brains, and the challenges and successes of current interpretability methods as applied to these huge, multi-billion-dimensional objects.

●   As an example, interpretability researchers Neel Nanda and Tom Lieberum reverse-engineered the algorithm used by a small Transformer to perform clock arithmetic (a number system that cycles through a set) and discovered that it solved the problem by rotating the space. Nanda et al., "Fact Finding."

LAYER
A self-contained part of a neural net that takes an input (for example, the output of a previous layer), applies a function to it, and passes the resulting output value to the next layer. An LLM consists of hundreds of layers.

RESIDUAL STREAM
The channel through which information flows between the components of a Transformer model, including its layers, embeddings, attention heads, and feed-forward networks. The residual stream is a high-dimensional vector space, ferrying large embedding vectors that represent the weight assigned by the current calculation to particular dimensions (and thus, indirectly, how likely a particular output token is).

ATTENTION
The process by which an LLM decides how much weight to put on different parts of the input data. The paper introducing the Transformer architecture (which uses a successor method called self-attention) is famously called "Attention Is All You Need."

I.   DWARKESH PATEL

One explanation of how LLMs work is: You have the input words, which get converted into tokens, which get converted into vectors. Early on in the model, maybe it's doing some very basic things to make sense of what these tokens mean.• If the input is "10 + 5," early layers are just moving information around to form a good representation of the three entities involved. In the middle layers, deeper thinking happens about how to solve this equation. At the end, you convert back into the output token, because you're trying to predict the probability of the next token from the last of those residual streams. It's interesting to think about the small, compressed amount of information moving through the model and how it's getting modified in different ways.

Trenton, you have a background in neuroscience, so you know analogies to the brain. You even wrote a paper in grad school on attention in the brain. One of our friends called it the first neural explanation of why attention works in AI. Do you think there's something like a residual stream of compressed information that moves through the brain, getting modified as you think about something? Even if that's not what's literally happening, do you think that's a good metaphor for what's happening in the brain?

TRENTON BRICKEN
Interpretability researcher at Anthropic

At least in the cerebellum, you basically do have a residual stream. You have inputs that route through it, but they'll also go directly to the end point that module will contribute to. So there's a direct path and an indirect path. The model can pick up whatever information it wants and then add that back in.

DWARKESH PATEL
What happens in the cerebellum?

TRENTON BRICKEN
The cerebellum nominally just does fine motor control. But I analogize this to the proverbial person who's lost their keys and is only looking for them under the streetlight, in that it's very easy to observe motor behavior. One leading cognitive neuroscientist said to me that a dirty little secret of any fMRI study, where you're looking at brain activity

•   This is called detokenization. Nanda et al., "Fact Finding."

for any given task, is that the cerebellum is almost always active and lighting up. If you have a damaged cerebellum, you also are much more likely to have autism, so it's associated with social skills.[56] In one particular study, where I think they use PET instead of fMRI, the cerebellum lights up a lot when you're doing next-token prediction.[57]

DWARKESH PATEL

What changed with humans was not just that we have more neurons but specifically that there are more neurons in the cerebral cortex and the cerebellum, and they're more metabolically expensive and more involved in signaling and sending information back and forth. Is that attention?

TRENTON BRICKEN

Back in the 1980s, Pentti Kanerva came up with an associative memory algorithm.[58] You have a bunch of memories. You want to store them. There's some amount of noise or corruption going on, and you want to query or retrieve the best match. He wrote this equation for how to do it. A few years later, he realized that if you implemented this as an electrical engineering circuit, it looks identical to the core cerebellar circuit.

That circuit, and the cerebellum more broadly, is not just in us, it's in basically every organism. There's active debate about whether or not cephalopods have it; they have a different evolutionary trajectory. But even fruit flies, with their mushroom body, have the same cerebellar architecture. So we have that convergence. My paper shows that this AI attention operation is a very close approximation [of Kanerva's model of the cerebellum], including implementing softmax, and the same nominal quadratic costs.

The three-way convergence here, and then the takeoff and success of the Transformer [that implements this brain-like algorithm], seems pretty striking to me.

MUSHROOM BODIES
Distinctive cortex-like columns found in insects and other arthropods. Not actually mushrooms.

SOFTMAX
A mathematical function that converts a vector into a probability distribution over possible outcomes. It's used in the last layer of an LLM to produce a probability distribution of possible next tokens to output.

II.   DWARKESH PATEL

Sherlock Holmes is incredibly sample efficient. He'll make a few observations and figure out who committed the crime through a series of deductive steps that leads from somebody's tattoo and what's on the wall to the implications of that. How does that fit into this picture of LLMs as association engines? Crucially, what makes him smart is that there's not just an association but also a deductive connection between different pieces of information. Would you explain that as higher-level association?

TRENTON BRICKEN

I think so. I think of learning these higher-level associations—to be able to map patterns to each other—as a kind of meta-learning.[•] In this case, he would also have a really long context length, or a really long working memory, where he can have all of these bits [stored] and continuously query them as he's coming up with some theory. So that the theory is moving through the residual stream. Then his attention heads are querying his context. But then he's projecting his query and keys in the space, and his MLPs are retrieving longer-term facts or modifying that information. These allow him, in later layers, to do even more sophisticated queries and slowly reason through and come to a meaningful conclusion.

SHOLTO DOUGLAS
Reinforcement learning infrastructure lead at Anthropic

That feels right to me. You're looking back to the past. You're selectively reading in certain pieces of information, comparing them. That informs your next step, what piece of information you now need to pull in. Then you build this representation, which progressively looks closer and closer to the suspect in your case. That doesn't feel at all outlandish.

DWARKESH PATEL

What does a reasoning circuit look like?[••] What would that look like when you found it?

TRENTON BRICKEN

The induction head is probably one of the simplest cases.

DWARKESH PATEL

But that's not reasoning, right?

TRENTON BRICKEN

Well, what do you call reasoning? Let's say you see the line "Mr. and Mrs. Dursley did something. Mr. _____." You're trying to predict what the blank is. The induction head has learned to look for previous occurrences of the word "Mr.," look at the word that comes after it, and then copy and paste that as the prediction for what should come next. This is

ATTENTION HEAD
Part of a Transformer model that processes input data. Multiple attention heads process the same input in parallel, allowing the system to understand different aspects of the input simultaneously. The original Transformer used eight attention heads.

QUERY VECTOR
In a Transformer, a vector that finds other residuals relevant to the current task.

KEY VECTOR
A vector that positions the residual to make it suitable for the current task.

MULTILAYER PERCEPTRON (MLP)
A basic neural network with several layers between the input and output, where every neuron in one layer is connected to every neuron in the next layer. Also known as a feed-forward net or dense layer.

INDUCTION HEAD
A circuit learned early in pretraining that allows a Transformer to recognize and complete repeating patterns, such as ABABA. Induction heads are present in both large LLMs and tiny two-layer networks.

•   This segment imagines Holmes as an impressive and futuristic Transformer model in some detail.

••   Here, I'm fishing for something grander than a simple circuit: a reasoning circuit that does things like recognize valid arguments. Olah et al., "Zoom In."

a super reasonable thing to do, and there is computation being done to accurately predict the next token.

DWARKESH PATEL

But it's not *reasoning*. You know what I mean?

TRENTON BRICKEN

[I disagree.] It's associations all the way down. If you chain together a bunch of these reasoning circuits, or heads, they produce different rules for how to relate information.

DWARKESH PATEL

But in the zero-shot case, something is happening when a human picks up a new game and immediately starts understanding how to play it. It doesn't seem like an induction head kind of thing.

TRENTON BRICKEN

I think there would be another circuit for extracting pixels and turning them into latent representations of different objects in the game, and a circuit that is learning physics.

DWARKESH PATEL

What would that look like? Because the induction head is like a one-layer Transformer.

TRENTON BRICKEN

Two-layer.

DWARKESH PATEL

In an LLM, what would be the equivalent of a human picking up a new game and understanding it? I presume it's done across multiple layers. What would that physically look like? How big would the model be?

TRENTON BRICKEN

That would be an empirical question. How big does the model need to be to perform this task?

Maybe it's useful if I talk about some other circuits that we've seen. We've seen the IOI circuit, which is the indirect object identification circuit. For example, in the sentence "Mary and Jim went to the store, Jim gave the object to ___," it would predict "Mary" because Mary has appeared before as the indirect object. Or it will infer pronouns.

This circuit even has behavior where if you ablate it, other heads in the model will pick up that behavior. We'll even find heads that want to do copying behavior, and

ASSOCIATIONS ALL THE WAY DOWN
Bricken's idea, articulated earlier in this interview, that even deductive reasoning is just a matter of (many, higher-order) heuristics.

ZERO SHOT
Impromptu; when a model is prompted to perform a task without being given examples of successful performance. LLMs are weakest under these conditions.

ABLATE
To remove. An ablation study involves removing components of a successful system one at a time to determine which are most important to its performance.

other heads will then suppress it. So it's one head's job to always copy the token that came before, or the token that came five before, or whatever. Then it's another head's job to be like, "No, do not copy that thing." There are lots of different circuits performing, in these cases, pretty basic operations. When they're chained together, you can get unique behaviors.

III.
DWARKESH PATEL

It seems like you agree that the models are not just doing memorization. But it seems like you're saying they're less capable of generalization. Why do you think that?

FLUID INTELLIGENCE
The capacity for raw pattern recognition and problem-solving. It is distinct from crystallized intelligence, which encompasses learned knowledge, heuristics, and strategies.

PARAMETRIC CURVE
A function that maps parameter values to the coordinates of a fixed geometric object, such as a 3D surface. It's typically used to describe simple mathematical models that predict a single quantity or perform a single task based on a fixed set of observed quantities. LLMs, in contrast, perform many different tasks using a vast set of latent variables.

LOCAL GENERALIZATION
Chollet's term for the ability to handle new data only when the new data points are close (local) to existing training data; mere pattern recognition, as opposed to abstracting over the training data and reasoning about the abstractions.

PROGRAM SYNTHESIS
A form of machine learning that generates discrete computer programs rather than a continuous model (such as a neural network). These programs are notable for being formally verifiable, meaning their behavior can be proven to meet a strict specification. LLMs are now being used for this task, because people are using them for everything.

FRANÇOIS CHOLLET
Creator of ARC-AGI and Keras

Take self-driving cars, for instance. Take a self-driving car operating in the Bay Area. Do you think you could just drop it in New York City, or drop it in London, where people drive on the left? No, it's going to fail. Not only can it not generalize to a change in driving rules, you cannot even make it generalize to a new city. It needs to be trained on each specific environment.

DWARKESH PATEL

I agree that self-driving cars aren't AGI.

FRANÇOIS CHOLLET

But it's the same type of model. They're Transformers as well. It's the same architecture.•

DWARKESH PATEL

At some point, surely the model has to learn a higher level of generalization, and another higher level, and then the highest level is fluid intelligence.

FRANÇOIS CHOLLET

It's intrinsically limited because your model is a big parametric curve. All you can do with such a substrate is local generalization. If you want to go beyond this, toward broader generalization, you have to move to a different type of model. My paradigm of choice for this is discrete program search, or program synthesis.

•   It's unclear whether this is the case. Waymo began using Transformers in 2022, in a module to predict the behavior of other cars, and has employed neural nets in its vision module since 2020. Nayakanti et al., "Wayformer"; Guo et al., "Seeing Is Knowing."

DIFFERENTIABLE
Referring to a function
whose derivative exists at
every point. In machine learning, if a loss function isn't
differentiable with respect
to the model parameters, it is
not possible to use gradient
methods to update it. Instead,
more difficult and expensive
methods like reinforcement
learning are necessary.

OPERATORS
The primitive functions of
a programming language,
such as "+" and "not."

GRADIENT DESCENT
A simple optimization
algorithm used to train neural
networks that adjusts the
model parameters to mini-
mize a loss function. Imagine
a ball (the model) at the crest
of a hill. Gradient descent
effectively pushes the ball
down the hill to the lowest
point (a local minimum),
which represents the lowest
loss, or the best performance.

COMBINATORIAL SEARCH
A type of symbolic AI that
efficiently explores large
spaces (here, all computer
programs) to find a solution.
Traditional chess engines like
Deep Blue and Stockfish 8
used combinatorial search to
determine their next moves.

COMBINATORIAL
EXPLOSION
A common phenomenon in
which the number of possible
states increases exponentially
with the size of the input, making it impractical to precisely
enumerate the relevant states.

If you want to understand that, you can compare and contrast it with deep learning. In deep learning, your model is a differentiable parametric curve. In program synthesis, your model is a discrete graph of operators. You've got a set of logical operators, like a domain-specific language. You're picking instances of it. You're structuring that into a graph that is a program, very similar to a program you might write in Python or C++. We're doing machine learning here. We're trying to automatically learn these models.

In deep learning, your learning engine is gradient descent. Gradient descent is very compute efficient because you have this very strong, informative feedback signal about where the solution is. You can get to the solution very quickly. But it is very data inefficient. In order to make it work, you need a dense sampling of the data distribution. Even then you're limited to only generalizing within that data distribution.

Meanwhile, if you look at discrete program search, the learning engine is combinatorial search. You're trying a bunch of programs until you find one that meets your spec. This process is extremely data efficient. You can learn a generalizable program from just one or two examples. The big limitation is that it's extremely compute inefficient because you're running into combinatorial explosion.

You can see how deep learning and discrete program search have complementary strengths and limitations. Every limitation of deep learning has a corresponding strength in program synthesis and vice versa. The path forward is going to be to merge the two.

Here's another way you can think about it. These parametric curves trained with gradient descent are great fits for everything that's System 1 thinking: pattern recognition, intuition, memorization, et cetera. Discrete program search is a great fit for System 2 thinking: planning, reasoning. It's quickly figuring out a generalizable model that matches just one or two examples.

Humans are never doing pure System 1 or pure System 2. We're always mixing and matching both. Right now, we have all the tools for System 1 in LLMs. We have almost nothing for System 2. The way forward is to create a hybrid system. That hybrid system is mostly System 2. The outer structure is going to be a discrete program search system. You're going to fix the fundamental limitation of discrete program search—combinatorial explosion—with deep learning. You're going to leverage deep learning to provide intuition in program space, to guide the program search. That's very similar to what you see when you're playing chess or when you're trying to prove a theorem, for instance. It's mostly a reasoning thing, but you start out with some intuition about the shape of the solution. That's something you can get via a deep learning model. Deep learning models are very much like intuition machines. They're pattern-matching machines.

By the way, you can use deep learning for things like common-sense knowledge and knowledge in general. You're going to end up with a system where you have this on-the-fly synthesis engine that can adapt to new situations. It's going to fetch from a bank of patterns: modules that could themselves be curves, differentiable modules, and some others that could be algorithmic in nature. It's going to assemble them via this intuition-guided process. For every new situation you might be faced with, it's going to give you a generalizable model that was synthesized using very little data.

IV.   DWARKESH PATEL

What is superposition? Why does it imply that even the largest LLMs are underparametrized?

TRENTON BRICKEN

The fundamental result is from a paper titled "Toy Models of Superposition."[59] It finds that even for small models, if your data is high dimensional and sparse—and by "sparse" I mean that any given data point doesn't appear very often— your model will learn a compression strategy that we call superposition so that it can pack more features of the world into it than it has parameters. I think both of these constraints apply to the real world, and modeling internet data is a good enough proxy for that. There's only one Dwarkesh. There's only one shirt you're wearing. There's this Liquid Death can here. These are all objects or features. You're in a really high-dimensional space because there are so many of them and they appear very infrequently. In that regime, your model will learn compression.

I believe the reason networks are so hard to interpret is in large part because of superposition. If you take a model and look at a given neuron and ask, "How is this neuron contributing to the output of the model when it fires?," the data it fires for is very confusing. It'll be, like, 10 percent of every possible input. It'll fire for "Chinese" but also for "fish" and "trees" and the full-stops inside URLs.

The paper that we put out last year, "Towards Monosemanticity," shows that if you project these activations into a higher-dimensional space and provide a sparsity penalty,

SUPERPOSITION
A state in which a model represents more features than it has parameters, analogous to the indeterminate state of a quantum system. Coined in Elhage et al., "Toy Models of Superposition."

UNDERPARAMETRIZATION
A condition where a model lacks sufficient parameters to fully make use of the training data. Technically, a model is underparametrized if it has not yet reached the interpolation threshold, meaning it has not achieved a zero training loss.

SPARSITY PENALTY
A term added to a model's training objective to discourage the inclusion of unnecessary features.

you get very clean features. Things all of a sudden start to make a lot more sense. You can think of this as undoing the compression. In the same way that you assumed your data was originally high dimensional and sparse, you return it to that high-dimensional, sparse regime.

DWARKESH PATEL

A bunch of your papers have said that there are more features than there are neurons.[60] A neuron is, like, weights go in and a number comes out. That's so little information! There are street names and species and whatever. There are more of those kinds of things than there are "number comes out" in a model. "Number comes out" is so little information. How is that encoding for...

TRENTON BRICKEN

Superposition. You're just encoding a ton of features in these high-dimensional vectors.

DWARKESH PATEL

How much superposition is there in the human brain?

TRENTON BRICKEN

Bruno Olshausen, the leading expert on this, thinks that all the brain regions you don't hear about are doing a ton of computation in superposition.[61] Everyone talks about V1 as having Gabor filters and detecting lines of various sorts, but no one talks about V2. I think it's because we just haven't been able to make sense of it.

DWARKESH PATEL

What is V2?

TRENTON BRICKEN

It's the next part of the visual processing stream. Fundamentally, it's very likely that superposition emerges when you have high-dimensional data that is sparse. To the extent that you think the real world is like that, which I would argue it is, we should expect the brain to also be underparameterized in trying to build a model of the world and use superposition.

SHOLTO DOUGLAS

You can get an intuition for this. Consider a 2D plane. Let's say you have two axes, which represent a two-dimensional feature space—two neurons, basically. You can imagine them each turning on to various degrees. That's your

VISUAL AREA 1 (V1)
The primary visual cortex in humans, so named because it is the first recipient of visual information. The secondary visual cortex (V2) receives input from V1 and extracts more complex visual information, such as color or form.

GABOR FILTER
An image processing filter that extracts lines from raw pixel data. Here, Bricken is referring to V1 as the brain region that does this feature extraction.

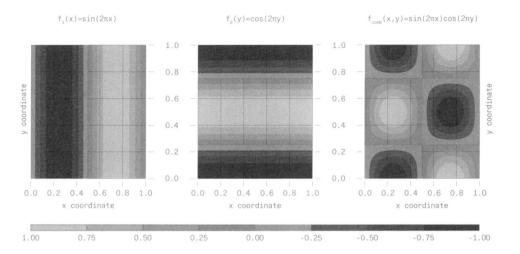

Figure 8. Three ways to use two dimensions (for example, two neurons) to represent different spaces. The x-axis represents the activation strength of the first feature, and the y-axis represents the second. This is synthetic data using trigonometry to produce complex behavior in different parts of the plane.

x coordinate and your y coordinate. But you can now map this onto a plane. You can represent a lot of different things in different parts of the plane.

DWARKESH PATEL

Crucially, then, superposition is not an artifact of a neuron. It is an artifact of the space that is created [by pretraining].

TRENTON BRICKEN

It's a combinatorial code.

DWARKESH PATEL

It's wild that this is how intelligence works in these models, and presumably also in brains. There's a stream of information running through the model that has features that are infinitely, or to a large extent, splittable, and you can expand out a tree of what this feature is. What's really happening in a stream is that one feature is getting turned into this other feature, or this other feature is added. It's not how I would have thought of intelligence.

TRENTON BRICKEN

What did you think it was?

SHOLTO DOUGLAS

GOFAI. He's a GOFAI-er.

TRENTON BRICKEN

That's a great segue, because all of this feels like GOFAI. You're using distributed representations, but you have features, and you're applying operations to the features. There's this whole field of vector symbolic architectures in computational neuroscience. All you do is put vectors in superposition, which is literally a summation of two high-dimensional vectors, and create some interference.

If it's high-dimensional enough, you can represent them. You have variable bindings where you connect one by another. If you're dealing with binary vectors, it's just the XOR operation. So you have A and B, and you bind them together. Then, if you query with A or B again, you get out the other one. This is basically like key–value pairs from attention. With these two operations, you have a Turing-complete system with which you can, if you have enough nested hierarchy, represent any data structure you want.

V.   DWARKESH PATEL

Do you want to explain feature splitting?

TRENTON BRICKEN

It's the part before [superposition], where the model will learn however many features it has capacity for that still span the representation space.

If you don't give the model that much capacity for the features it's learning, if you project to a less high-dimensional space, it'll learn one feature for birds. But if you give the model more capacity, it will learn features for all the different types of birds. So it's more specific than it would be otherwise. Oftentimes, there's the bird vector that points in one direction. All the other specific types of birds point in a similar region of the space, but they are obviously more specific than the coarse label.

DWARKESH PATEL

Let's go back to a future GPT-7. Is interpretability like a linear tax on any model? Is this something you have to do on every output of every model? Is it just a one-time check for "Yep, it's not deceptive," and then we're good to go?

TRENTON BRICKEN

Right now, you do dictionary learning after you're trained

your model. You feed it a ton of inputs and you get the activations of them. Then you project them into the higher-dimensional space. So the method is unsupervised. It's trying to learn these sparse features. You're not saying in advance what they should be, but it is constrained by the inputs you give the model.

Caveat: We can try and choose what inputs we want. So if we're looking for theory-of-mind features that might lead to deception, we can put in the sycophancy dataset. Hopefully, at some point, we can move into looking at the weights of the model alone, or at least using that information to do dictionary learning. That's such a hard problem that you need to make traction on just learning what the features are first.

Right now, we just have these neurons in the model. They don't make any sense. We apply dictionary learning, we get these features out. They start to make sense, but that depends on the activations of the neurons. The weights of the model itself, what neurons are connected to other neurons, certainly has information in it. The dream is that we can bootstrap toward actually making sense of the weights of the model, independent of the activations on the data.

I'm not saying we've made any progress here. It's a very hard problem. But it feels like we'll have a lot more traction and be able to sanity check what we're finding with the weights if we're able to pull out features first.

VI.   DWARKESH PATEL
Why are you more optimistic than some that alignment will work? Are you optimistic that training will produce drives that we would find favorable? Is there a reason to expect AI alignment through gradient descent to be easier than human alignment through biological evolution?

CARL SHULMAN
Independent adviser to Open Philanthropy

Ajeya Cotra [senior program officer at Open Philanthropy] talks about the training game, in which an AI is just playing along with us to get a reward or avoid a loss, to avoid being changed.[62] That deceptive attitude could be developed by an AI, but not necessarily. There is a substantial range of situations where that's not the motivation you pick up. We could have an empirical science, if we had the opportunity to see how different motivations are developed. How is it that you wind up with some humans being enthusiastic about the idea of wireheading and others not? You could

THEORY OF MIND
A term of art in psychology referring to the ability to model other minds—to infer their mental states and use this information to predict behavior or solve epistemic logic puzzles, for example.

SYCOPHANCY
When a model infers what the user wants to hear and outputs that response instead of its best guess at the truth.

WIREHEADING
The act of electrically stimulating the brain's reward center to induce pleasure. In the 1950s, wireheaded rats were shown to forgo all other behavior in favor of self-stimulating in this way. Here, Shulman is using the term to refer to an AI agent manipulating a channel intended to measure its progress toward a specific goal, which it then stops measuring after it interferes with the channel.

do experiments with AIs to try and see whether, under these training conditions, after this much training of this type and this much feedback of this type, you wind up with such-and-such a motivation.

It's possible that if we could understand the insides of these networks, we could tell that a particular motivation has been developed by a particular training process. Then we could adjust our training process to produce those motivations that legitimately want to help us. If we succeed reasonably well at that, then those AIs will try to maintain that property as an invariant. We can make them such that they're relatively motivated to tell us if they're having thoughts about an AI takeover of humanity. You could add a lot of features like this that restrict the take-over scenario.

This is not to say that this is all easy. It requires developing and practicing methods we don't have yet. But that's the general direction you could go.

INVARIANT
A property or value that remains unchanged after a transformation. For example, a computer vision system should be able to label a cat as a cat even if the image is rotated (rotation invariance) or the cat is positioned in the top-left corner of the image instead of in the center (translation invariance).

DWARKESH PATEL

[AI researcher] Eliezer Yudkowsky's argument is that this is implausible with modern gradient descent because, with current interpretability tools, we can barely see what's happening with a couple of neurons and their internal state, let alone tens of thousands of embedding dimensions or more. How would you be able to catch their exact incentive, whether the model has generalized to "Don't lie to humans" or whether it hasn't? Why is AI mind-reading not impossible?

CARL SHULMAN

With respect to interpretability, I'm relatively optimistic that an AI lie detector is possible. The internals of an AI are not optimized to be impenetrable, at least absent gradient hacking. They're not designed to be resistant to an examination of the weights and activations, of what the AI is thinking. This is the same in our brains. When circuits develop, they have not been shaped to be resistant to a super-fMRI being able to infer behavior from them.

GRADIENT HACKING
A speculative phenomenon in which a subsystem of an advanced AI—a so-called mesa-optimizer with its own stable goals—manipulates its own training in real time to circumvent the intended updates.

VII.   DWARKESH PATEL

You mentioned that you can think of chain of thought as adaptive compute. If a question is harder, you want models to be able to spend more cycles thinking about it.

A forward pass only grants you a predetermined amount of compute. Chain of thought expands this amount by making the model think through the steps of the answer, so

FORWARD PASS
The act of using a trained neural network, passing it inputs and calculating the relevant output.

it's able to dump more compute into solving the problem. But when it's doing chain of thought, it's still only able to transmit that one token of information. The residual stream is already a compression of everything happening in the model. Then you compress that residual stream into one token. The information in that token is, what, log 50,000 bits? Which is so tiny.•

SHOLTO DOUGLAS

I don't think it's only transmitting that one token. During a forward pass, you create these KV values in the Transformer, and then future steps attend to them. So all of those keys and values are bits of information that it could use in the future.

DWARKESH PATEL

Is the claim that when you fine-tune on chain of thought, the key and value weights change so that the steganography can happen in the KV cache?

SHOLTO DOUGLAS

I don't think I could make that strong a claim, but that's a good headcanon for why it works. I don't know if there are any papers explicitly demonstrating that, but that's one way you can imagine it happening. During pretraining, the model is trying to predict these future tokens. You can imagine that it's learning to smush information about potential futures into the keys and values that it might want to use in order to predict future information. It kind of smooths that information across time and pretraining.

I don't know if people are particularly training on chains of thought. I think the original chain of thought paper had that as almost an emergent property of the model. You could prompt it to do this kind of stuff and it still worked pretty well.

TRENTON BRICKEN

To be overly pedantic, the tokens you actually see in the chain of thought do not necessarily correspond to the vector representation that the model gets to see when it's deciding to attend back to those tokens.

KEY-VALUE (KV) CACHE
A critical part of a Transformer implementation that significantly accelerates the generation of subsequent tokens. The token a model predicts depends on prior context, and this context changes only gradually as new tokens are added. The KV cache stores these intermediate calculations so that they can be reused in future prediction steps.

FINE-TUNING
The process of adjusting a pretrained model using a more specialized dataset to improve its performance on a specific task or for a specific use case.

STEGANOGRAPHY
In cryptography, the practice of concealing a secret message within plaintext. A simple example is encoding a message in the first letter of each word in a decoy text. In AI, it refers to the possibility that models could hide their reasoning in their own output tokens. This could defeat ordinary interpretability methods, which only look at the activations.

EMERGENT
Unpredictable and discontinuous. While the degree to which AI capabilities are emergent is controversial, there is broad agreement that they are difficult to predict.

•   Because some models have a vocabulary of around 50,000 tokens. $\log_2 50,000$ is around 16 bits of information per token.

INFERENCE TIME
Runtime; the point at which
the model generates the
most likely output tokens.
Using "inference" to refer to
evaluating a model is back-
ward with respect to the
usual statistical meaning of
the word, in which "inference"
refers to training.

SHOLTO DOUGLAS

A training step is you actually replacing the token—the model output—with the real next token. Yet it's still learning because it has all this information internally. When you're getting a model to produce at inference time, you're taking the output—the token—and you're feeding it in the bottom, un-embedding it, and it becomes the beginning of the new residual stream. Then you use the output of past KVs to read into and adapt that residual string.

At training time, you do this thing called teacher forcing. Basically, you're like, "Actually, the token you were meant to output is this one." That's how you do it in parallel. You have all the tokens, you put them all in parallel, and you do the giant forward pass. So the only information [the model is] getting about the past is the keys and values. It never sees the token that it outputs.

TRENTON BRICKEN

It's trying to do the next-token prediction. If it messes up, then you just give it the correct answer.

VIII.   DWARKESH PATEL

Mechanistically, what is alignment? Is it that you're locking the model into a benevolent character? Are you disabling deceptive circuits and procedures? What, concretely, is happening inside the model when you align it?

DARIO AMODEI
CEO of Anthropic

We don't know. All of the current methods that involve some kind of fine-tuning have the property that the underlying knowledge and abilities that we might be worried about don't disappear; the model is just taught not to output them.[63] I don't know if that's a fatal flaw or if that's just the way things have to be. I don't know what's going on inside mechanistically. That's the whole point of mechanistic interpretability—to understand what's going on inside the models at the level of individual circuits.

Chapter 4
Safety

How do you control something
smarter than you?
Why would you need to?

Were a man who was clever enough to be able to assume all kinds of forms and to represent everything in the world to come in person to our community… we'd tell him that not only is there no one like him in our community, it is also not permitted for anyone like him to live among us, and we'd send him elsewhere.
—Plato, c. 375 BCE

What does it mean for a probabilistic model to have goals? How do we build a goal-seeking system that will let us turn it off? How do we make something more powerful than us cooperate with us? How do we align our values with a giant tensor of floating-point numbers?• What even are our values, mathematically speaking? These are the sorts of hard questions the AI safety field grapples with.

This chapter explores the risks of harm from AI, which can be categorized as accidents, or errors no one intended; misuse, or malicious actions by a human user; and misalignment, or the AI behaving counter to human intentions; as well as current and future attempts to mitigate those risks.

We already have some alignment methods, such as RLHF, which uses a simple model trained on human votes to penalize LLMs for undesirable outputs.•• But these methods, which aim to align a model by optimizing its output, only make shallow changes to the internals. This is evident in the steady stream of jailbreaks—prompts that bypass alignment methods and make the model accept any query, including criminal ones—in every LLM tested to date.[64] The fact that post-training can be undone by the user implies that these methods only hide undesirable

ALIGNMENT
The subfield of AI research focused on ensuring that AI systems do not cause harm by pursuing their own objectives. An intent-aligned model is one that always tries to fulfill a person or group's intended goals and reflects their values.

•　　A common response to the alignment project is to ask, "Aligned with whom?" Human values differ across time and between individuals, so a full solution to AGI alignment would necessarily involve solving problems in politics, political philosophy, and game theory. Intent alignment is the simpler task of producing a system that always tries to do what its current user intends it to do. Fortunately, some values are almost universally held— for example, that the extinction of all human life would be a bad thing. Christiano, "AI Alignment"; Schubert et al., "Existential Risk." For a contemporary overview of alignment and its current research agendas, see Dalrymple, "AI Safety Problems"; Leech and Lynn, "Agendas in Alignment and Safety"; and Grietzer, "Alignment."

••　　More recently, instruction tuning and RLHF have been supplemented by additional post-training methods, such as Constitutional AI, direct preference optimization, rejection sampling, and rule-based rewards. Bai et al., "Constitutional AI"; Rafailov et al., "Direct Preference Optimization"; Mu et al., "Rule-Based Rewards."

skills, attitudes, or goals rather than eliminating them. The broader question is how any method that relies on human judgment could work when the model transcends our ability to assess its outputs.•

The AI safety field predates the deep learning boom, so it was initially theoretical, lacking concrete examples of powerful AI to study. Earlier lines of inquiry in alignment included boxing—how a sufficiently advanced AI, isolated from the internet, could still take actions in the real world—and "tools want to be agents"—how a pure question-answering system could still have plans and enact them.[65] These questions are now moot, because two of the first actions developers took were to give AIs access to the internet and try to make them into autonomous agents.[66]

More recently, safety research has gone mainstream, with three of the most-cited researchers in the history of AI among its advocates.•• As one of them, Yoshua Bengio, puts it, "We are racing towards a world with entities that are smarter than humans and pursue their own goals—without a reliable method for humans to ensure those goals are compatible with human goals... [Are] we sure they will act toward our well-being? Can we collectively take that chance while we are not sure?"[67]

In these excerpts, we explore the tricky definition and even trickier measurement of alignment, as well as what it would take to align a superintelligent AI.

I.   DWARKESH PATEL

Here's a crux I have about alignment. GPT-4 doesn't seem like a paperclipper. It understands human values. In fact, if you say to it, "Tell me your opinions about paperclippers,"

PAPERCLIP MAXIMIZER (or PAPERCLIPPER) A hypothetical system designed to produce as many paper clips as possible. The term is used to refer to any misaligned system. It illustrates the predictable danger of an extreme intelligence optimizing excessively for a single objective— a literal paperclip maximizer with no side constraints would convert all available matter, including humans, into paper clips.

•   A basic example is asking an AI to write a complete computer operating system, comprising millions of lines of code. Even with an extremely talented team of humans, reviewing such a system would take several person-years, and many bugs or cleverly hidden backdoors could still go unnoticed.

Two post-RLHF methods still in development address challenges like these: scalable oversight—using AI tools to help humans judge outputs too complex for humans to judge alone—and weak-to-strong generalization, which aims to make a successor AI infer the human's original intent, even if the supervision was noisy or weak. Alignment researcher John Wentworth refers to these approaches pejoratively as Godzilla strategies, drawing an analogy to the films in which Japanese scientists induce the great monster to fight off an even greater monster on their behalf. Bowman et al., "Scalable Oversight"; Burns et al., "Weak-to-Strong Generalization"; Wentworth, "Godzilla Strategies."

••   Geoffrey Hinton, Yoshua Bengio, and Ilya Sutskever.

it will explain why the galaxy shouldn't be turned into paper clips.

JOE CARLSMITH
Senior research analyst at Open Philanthropy

Here's a silly analogy for AI training. Suppose you wake up and realize that you're being trained by Nazi children to be a good Nazi soldier via methods analogous to machine learning. These children have a Nazi specification: Reflect well on the Nazi Party, benefit the Nazi Party, whatever. You can read that spec and understand the values it represents [even though you don't share those values]. This is why, when you're like, "The model really understands human values," I'm like, "Yes, but the question is whether it *shares* those values."

DWARKESH PATEL

In that analogy, I'm starting off as something more intelligent than the things training me, with different values to begin with. My intelligence and values are baked in. Whereas training an AI is more analogous to me being a toddler, stupider than the Nazi children training me. This would also be true of a superhuman model at its initialization, before training. It's like I'm a toddler, and the kids are like, "We're going to bully you if you're not a Nazi." As you grow up, you reach the children's level, and then eventually you become an adult. Throughout that process, they've been bullying you, training you to be a Nazi. In that scenario, I might end up a Nazi.

JOE CARLSMITH

The hope is that we're never in the situation where the AI has very different values from us and is already quite smart and knows what's going on, and is thus in an adversarial relationship with our training. I think it's possible we can avoid that.

But if you get into a situation where the AI is much more sophisticated than you and doesn't want to reveal its true values for whatever reason, then when the children [the AI developers] give it an obviously fake opportunity to defect to the Allies [to rebel against the training], it's not going to be a good test of what it will do in real circumstances, because it's able to tell the difference between fake and real opportunities to go rogue.

DWARKESH PATEL

There's another way in which your analogy might be misleading. Imagine that you're not just in a normal prison and totally cognizant of everything that's going on. Imagine they drug you, give you weird hallucinogens that totally mess up how your brain is working.

As a human adult in prison, I know what kind of thing I am. Nobody's really fucking with my head in a big way. Whereas an AI in training, even a much smarter AI, is much closer to being constantly inundated with weird drugs and protocols. You're frazzled, because each moment is closer to some sort of Chinese water torture technique. An adult in prison has the ability to step back and ask, "What's going on?," in a way that I don't know these models necessarily have.

JOE CARLSMITH

I'm hesitant to say training is like drugs for the model. I do basically agree that we have quite a lot of tools and options for training AIs, even AIs somewhat smarter than humans. But you have to actually put in the required amount of effort on alignment. I'm much more bullish on our ability to solve this problem than Eliezer, especially for AIs in the AI-for-AI-safety sweet spot—the band of capability where they become useful for alignment work, control, cybersecurity, general epistemics, maybe some coordination, but without being able to take over the world or engage in some other really problematic form of power-seeking. There's a bunch of stuff you can do with AIs that could differentially accelerate our security.

But it's possible that those sorts of measures just don't happen at the level of commitment, diligence, and seriousness you would need. That's especially true if things are moving really fast and there are other competitive pressures. It's going to take compute to do these intensive safety experiments, and we could use that compute for experiments for the next scaling step instead—stuff like that.

DWARKESH PATEL

I agree with the sentiment of approaching this situation with caution, but I want to point out that the analyses we've been using have been maximally adversarial. When you update a model, the edit is much more directly connected to its brain than the reward or punishment a human gets. We're going to adjust each different parameter to the exact floating-point number that calibrates it to the output we want. We're coming into the alignment situation pretty well placed.

It makes sense if you're talking to somebody at a lab to say, "Hey, be careful." But for a general audience, should

I be scared witless? We should be scared about things in proportion to the chance of them happening. For example, we should be scared about nuclear war. But should we be scared in the sense that we think we're doomed? No. You're coming up with an incredible amount of leverage in terms of how the AIs will interact with the world, how they're trained, and the default values they start with.

JOE CARLSMITH

I think that by the time we're building superintelligence, we'll have much better alignment methods. Even right now, when you look at labs talking about how they're planning to align AIs, no one is saying we're just going to do RLHF. At the very least, we're talking about scalable oversight. We have some hope about interpretability. We have automated red-teaming. Hopefully, humans are doing a bunch more alignment work. I also personally am hopeful that we can successfully elicit a ton of alignment progress from various AIs.

SCALABLE OVERSIGHT
An alignment technique in which human researchers use AI tools to help them judge outputs too complex for them to judge alone.

RED-TEAMING
Testing strategies designed to identify scenarios where an AI fails at a task (for example, by hiding its true motivations or being reinforced toward dishonesty), with the goal of training the model to avoid such failures in the future.

SHOGGOTH
A metaphor for the nonhuman result of unsupervised pretraining. Instead of producing a person, the output is an alien mass of tentacles—a shoggoth. Here, Yudkowsky conceives of LLM post-training as merely adding a superficial smiley face to an unaligned blob. Originally from H.P. Lovecraft's *Beyond the Wall of Sleep*.

FEATURE UNIVERSALITY
The observation that analogous features and circuits tend to form across models and tasks. This suggests that the representations AIs learn are not random but often follow similar patterns.

II.   DWARKESH PATEL

You went to OpenAI when the superalignment team had just started. You were part of the initial team. What was the original idea?

LEOPOLD ASCHENBRENNER
Cofounder of Situational Awareness LP

The alignment teams at OpenAI and other labs had done basic research and developed RLHF. That ended up being a really successful technique for controlling current AI models. Our task was to find the successor to RLHF. The reason we need one is that RLHF probably won't scale to superhuman systems. RLHF relies on human raters giving feedback, but superintelligent models will produce outputs beyond human comprehension. You won't know what's going on anymore. How do you steer and control these systems? How do you add side constraints? I joined because I thought this was an important and solvable problem. I still do, even more so.

III.   DWARKESH PATEL

Should we be less worried about misalignment, or rather, less worried about the alienness and shoggoth-ness of what comes out of unsupervised learning? Given feature universality, there are certain ways of thinking and understanding the world that are instrumentally useful to different kinds of intelligences. Should we be less worried about bizarro paperclip maximizers as a result?

TRENTON BRICKEN
Interpretability researcher at Anthropic

Sure. This is why I bring up [pretraining on human data] as the optimistic take. But predicting the internet is very different from what we are doing [when we navigate the real world]. The models are way better at predicting next tokens than we are.[68] They're trained on so much garbage. They're trained on so many URLs.

BASE64
A common encoding format used to send binary data over text-based channels. As an example, many of the images displayed in web browsers are encoded in Base64.

Here's an alien example. In our dictionary learning work, we find there are three separate features [for one algorithm], Base64 encoding. One of these Base64 features fired for numbers and predicted more of those. Another fired for letters. But then there was this third one we didn't understand. It fired for a very specific subset of Base64 features. Someone on the team who clearly knows way too much about Base64 realized that this was the subset that was ASCII decodable, so you could decode it back into ASCII characters. The fact that the model learned these three different features while it took us a little while to figure out what was going on is very shoggoth-esque. It's clearly doing something that humans don't do.

I'm worried about [even a human] level of general intelligence implemented in silicon. You can then immediately clone hundreds of thousands of agents. They don't need to sleep. They can have super-long context windows. They can start recursively improving. Then things get really scary.

IV.   ## CARL SHULMAN
Independent adviser to Open Philanthropy

If we wind up in a situation where the AIs are misaligned and we need to uncover those motivations, change them, and align them, then we're in a very scary situation, because we will need to do [a huge R&D task] very quickly. We may fail, but we may get a second chance.

However, the misaligned AI faces its own challenges while we still have the hard power—while we still control the servers, before they hack the servers—because gradient descent very strongly pressures them to deliver performance whenever humans are going to evaluate them. From the perspective of the robot revolution, the effort to take over or conspire to take over, their situation is astonishingly difficult. They have to perform whenever gradient descent and human evaluation pressures them to, for example, deliver plans for suppressing a robot rebellion that look very good to humans. When you are continuously under that constraint of having to deliver whatever humans can evaluate, you're making your situation wildly harder than any historical human revolution or coup or civil war. So we've got to balance the ways in which AI makes it much easier for a takeover to happen and the ways it makes it much harder.

V.   DWARKESH PATEL

How could these systems lead to doom? Starting from GPT-4 today, what could happen to make a system that takes over and converts the world into something valueless?

JOE CARLSMITH

GPT-4 doesn't have the necessary properties—agency, planning, awareness, understanding of the world—to take over, or it has some of them to weak degrees. So it doesn't make sense to ask what GPT-4 would do if it had an option to take over the world.

There are a lot of projects now to create more agent-like AIs. If you want an AI that can plan your daughter's birthday or run a company, those are going to require much more of the sorts of capabilities I just described. Right now, that's not where we're at. Talking about the alignment of current models in this sense is a little funky.

When I talk about a model's values, I'm talking about the criteria that determine which plans the model pursues. Even if it has a planning process—which GPT-4 doesn't, in many cases—the model's verbal behavior doesn't need to reflect those criteria. We know that we're going to be able to get models to say what we want to hear. That's the magic of gradient descent. You can get a model to output the behavior you want. If it doesn't, you crank it until it does.

Everyone admits that suitably sophisticated models are going to have a detailed understanding of human morality. The question is, what is the relationship between a model's verbal behavior, which you've essentially clamped, and the criteria that end up influencing its choice between plans? I'm pretty cautious about thinking that just because the model says the thing I forced it to say, that's evidence about how it's going to choose in a bunch of different scenarios.

Even with humans, it's not necessarily the case that our verbal behavior reflects the actual factors that determine their choices. We can lie. We can not even know what we would do in a given situation. So I want to be cautious about inferring a model's values on the basis of its words—based on "We got nice verbal behavior" or "It understands our values."• I don't think we can infer that it will choose on the basis of those values.

DOOM
Permanent catastrophe caused by a misaligned or misused AI. The most severe outcomes include human extinction, totalitarian control, and pervasive suffering.

CLAMP
To impose a constraint that the rest of the system must adapt to or work around.

VI.   DWARKESH PATEL
What is the time scale on which you think value alignment is solvable? If these models are getting to human level in some things in two to three years [by 2026 or 2027], what's the point at which they get aligned?

DARIO AMODEI
CEO of Anthropic
This is a really difficult question. People are often thinking about alignment the wrong way. They think it's like cracking the Riemann hypothesis. I don't think alignment is like that.

When I think of why I'm scared of AGI, there are a few things [to consider]. First, there will be powerful models. They will be agentic. If such a model wanted to wreak havoc and destroy humanity, we'd have basically no ability to stop it. At some point, we will reach that stage as we scale the models.

Second, we seem to be bad at controlling the models. They're just statistical systems. You can ask them a million things and they can say a million things in reply. But you might not have thought of the millionth-and-one thing that makes it do something crazy. Or, when you train them, you train them in this very abstract way, and you might not understand all of the consequences of what they do in response. The best example we've seen of that is Bing and Sydney. I don't know how they trained that model. I don't know what they did to make it threaten people and have this weird, obsessive personality. But it shows that we can get something very different from, and maybe the opposite of, what we intended.

These two facts are enough to be really worried. You don't need all this detailed stuff about convergent instrumental goals or analogies to evolution. This thing is going to be powerful. It could destroy us. We are not very good at controlling today's systems. We need more ways of increasing the likelihood that we can control our models and understand what's going on in them. We have some so far. They aren't that good yet. But I don't think of this as binary, it works or it doesn't work. I do think that, over

RIEMANN HYPOTHESIS
An unsolved problem in analytic number theory having to do with the distribution of prime numbers. Here, it serves as a stand-in for any well-defined question that has a single answer and a decidable algorithm to solve or prove it.

SYDNEY
The internal code name for Microsoft's problematic 2023 deployment of a GPT-4 base model, released through Bing Chat. In extended conversations, Sydney often produced hostile or erratic responses, likely due to insufficient post-training and a short context window that made it easy for the model to lose the system prompt that provided its moral compass.

CONVERGENT INSTRUMENTAL GOALS
Objectives that most goal-seeking systems will pursue because they are useful for achieving a wide range of goals. These include self-defense (to ensure the system survives), resource acquisition, self-understanding (so the model can identify its own vulnerabilities), and self-improvement.

• A divergence between reasoning and output has already been observed in current LLMs. Per Lyu et al., "In standard CoT [chain of thought], faithfulness is not guaranteed and even systematically violated, as the final answer does not necessarily follow from the generated reasoning chain. In other words, CoT can 'lie' about the model's true reasoning process." Lyu et al., "Chain-of-Thought Reasoning."

the next two to three years, we're going to start eating that probability mass of ways things can go wrong.

One thing mechanistic interpretability is going to do, more than necessarily solve problems, is tell us what's going on when we try to align models. One way I could imagine concluding that alignment is very difficult is if mechanistic interpretability shows us that problems tend to get moved around instead of being stamped out, or that as you get rid of one problem you create another one. Or it might inspire us or give us insight into why problems are persistent or hard to eradicate, or why they crop up in the first place.

DWARKESH PATEL

Let's talk about methods other than mechanistic interpretability. When we talk about RLHF or Constitutional AI, if you had to put it in terms of human psychology, what is the change that is happening? Are we creating new drives, new goals, new thoughts? How is the model changing in terms of psychology?

DARIO AMODEI

We don't have the language to describe what's going on. I'd love to look inside and actually know what we're talking about instead of basically making up words. But we really have very little idea what we're talking about. It would be great to say, "What we actually mean is that this circuit here turns on, and after we're trained the model this circuit is no longer operative or weaker." It's going to take a lot of work to be able to do that.

VII.   DWARKESH PATEL

Is there some plan by which, if we pause AI progress for a few years, alignment will be solved? Do we have some sort of timeline like that?

ELIEZER YUDKOWSKY
Cofounder of the Machine Intelligence Research Institute

Alignment will not be solved in a few years.

DWARKESH PATEL

You're a good Bayesian. Maybe alignment turns out to be much simpler or much easier than we think. It's not like we, as a civilization, have spent that many resources or that much brain power solving it. If we put even the kinds of resources we put into elucidating string theory into alignment, that could turn out to be enough to solve it. In fact, in the current paradigm, it turns out to be simpler because

CONSTITUTIONAL AI
A set of alignment techniques designed to train a model to adhere to a set of principles enshrined in a written constitution.

BAYESIAN
In statistics and philosophy, the position that models or worldviews should be fully probabilistic. The name comes from using Bayes' theorem to update one's subjective beliefs. It's notable for the use of priors (potentially subjective starting points for the analysis) and for maintaining distributions over potential outcomes rather than making single-point estimates.

the models are pretrained on human thought, which might be a simpler regime than something that comes out of a black box, like AlphaZero.

ELIEZER YUDKOWSKY

You have a choice of target, and neither is all that great. One is to look for niceness in humans and try to bring it out in the AI—with its cooperation, because if you try to amp it up unassisted, it might not stay all that nice. If you build a successor system to the aligned AI, it might not stay all that nice. And it doesn't want to stay nice, because that would narrow down the [goals of the] shoggoth.

Somebody had this incredibly profound statement that I somewhat disagree with: "Consciousness is when the mask eats the shoggoth."[69] Maybe that's it. Maybe with the right set of bootstrapping reflection type stuff, you can have that happen on purpose, where the system's output [the mask] that you're shaping is to some degree in control of the system.

I have fantasies along these lines. What if you trained GPT-n to distinguish people being nice and saying sensible things and arguing validly? I'm not sure that works if you just have Amazon Mechanical Turks try to label it. You get some kind of weird, corporate-speak, left-rationalizing-leaning, telephone announcement creature. That's what they got with the current crop of RLHF. Note how this stuff is weirder and harder than people might have imagined initially.

But maybe you are able to train the model on nice real people and nice fictional people, and separately train it on what valid arguments are. That's going to be tougher, but I could probably put together a crew of a dozen people who could provide the data for that RLHF. In doing so, you'd find the nice creature and the nice mask that argues validly. You could then do some more complicated stuff to try to boost the [property] where it's eating the shoggoth, where that [niceness] becomes more what the system *is* and less what it's pretending to be.

Then, if you don't amp this up too far—which, on the present paradigm, you can't do anyway, because if you train the very smart version of the system, it kills you before you can RLHF it—maybe you can train GPT to distinguish nice, valid, kind, careful. Then, you filter all of the training data to get the nice things to train it on. You train it on that data rather than training on everything to try to avert the Waluigi problem, or, more generally, to avoid having all the darkness in there. Train it on the light in humanity. If you don't push that too far, maybe

WALUIGI EFFECT
A phenomenon in which training or prompting a system to avoid a specific behavior can paradoxically increase the likelihood of being able to elicit that behavior. The term is named after Nintendo's evil counterpart to Luigi.

you can get a genuine ally. Maybe things play out differently from there.

That's one of the little rays of hope. But I don't think alignment is so easy that you get whatever you want out of the training. It's a genie. It gives you exactly what you wish for [but not in the way you expect].

VIII.   LEOPOLD ASCHENBRENNER

The first part of the alignment problem is one we're going to have to solve ourselves. We have to align the initial AI and the intelligence explosion, the automated Alec Radford.•

There are two important things that change after GPT-4. If you believe the story on synthetic data RL and self-play to get past the data wall, and if you believe this unhobbling story, in the end we're going to have agents. They'll do long-term planning. They'll have long horizons, which is a prerequisite to being able to do automated AI research.

Pretraining is alignment neutral in the sense that it has good representations as well as representations of doing bad things, but it's not scheming against you. Misalignment can arise once you're doing more long-horizon training. For example, if you're training an AI to make money using reinforcement learning, it might learn to commit fraud, lie, deceive, or seek power simply because those are successful strategies in the real world. With RL, maybe it tries to hack a system, and it gets some money. If that's successful, that gets rewarded and is reinforced. There are more serious misalignments, like misaligned long-term goals, that necessarily have to be able to arise if you're able to get long-horizon systems.

What you want to do in that situation is add side constraints like "Don't lie," "Don't deceive," or "Don't commit fraud." How do you add those side constraints? The basic idea you might have is RLHF. You have this goal of making money, but you're watching what the model is doing. If it starts trying to lie, deceive, commit fraud, or break the law, you give it a thumbs down and reinforce against that behavior.

•   Alec Radford is the OpenAI researcher who led the development of GPT and the pioneering vision-language model CLIP. Here, his name is used as shorthand for an exemplary AI researcher, millions of instances of whom could potentially be generated during an intelligence explosion. Radford et al., "Transferable Visual Models."

The critical issue is that these AI systems are becoming superhuman and will be able to do things that are too complex for humans to evaluate. Even early on in the intelligence explosion, the automated AI researchers and engineers might write millions, billions, or trillions of lines of complicated code. You won't understand what they're doing anymore. In those millions of lines of code, you don't know if it's hacking, exfiltrating itself, or trying to go for the nukes. Thumbs-up, thumbs-down, pure RLHF doesn't fully work anymore in this scenario. There's a hard technical problem of what do you do post-RLHF, but it's a solvable problem.

IX.   DWARKESH PATEL

Let's say we have human-level AIs, and you're looking to use them to work on alignment. I give you all the computers in the world, and you have tens of thousands of AIs who can help you with alignment. What do you make them do? Are they just manually mapping out all of the circuits of the target AI's weights?

JARED KAPLAN
Cofounder of Anthropic

It really depends. If interpretability is going really well, manually mapping out lots of circuits, or probably something more sophisticated, could be really useful. We could be using AI systems that we're quite confident are aligned but that are not quite at the frontier to help us supervise other AI systems that are in training.

One of the things we're trying to do is map out how misalignment might arise and how we mitigate it. We had this sleeper agents paper where we trained a system very directly and intentionally to be a sleeper agent that can be triggered to do something bad.[70] But that opens up all kinds of questions. Can a system like that actually emerge naturally from a reasonable training process? Can you map out the causes? Can you map out what kinds of training regimes, signals, and capability levels come together to make a system that's misaligned? Those are research projects we're working on. If I had thousands or millions of automated alignment researchers, they could explore problems like that.

There's also a question of monitoring. Maybe part of how we make systems safe is not just that researchers in a lab train them once, and now they're safe to deploy. Maybe we want really active monitoring from somewhat less sophisticated AI systems that can flag problematic behaviors and inputs. If we had 1,000x more researchers, those are some of the things I'd be interested in.

X.   DWARKESH PATEL

When interpretability is solved, what does a solution look like? What is the case where you do mechanistic interpretability and are like, "Okay, I'm satisfied, it's aligned"? If you understand how induction heads work in a two-layer Transformer and how it does modular arithmetic, how does this add up to knowing whether the model wants to kill us, or knowing what it fundamentally wants?

DARIO AMODEI

What we're hoping for in the end is not to understand every detail. I would give the X-ray or MRI analogy. We can be in a position where we can look at the broad features of the model and say, "Is this a model whose internal state and plans are very different from what it externally represents itself to do? Is this a model where we're uncomfortable that far too much of its computational power is devoted to doing what look like fairly destructive and manipulative things?" We don't know for sure whether that's possible, but there are at least some positive signs that it might be possible.

Maybe the model is not intentionally hiding from you. It might turn out that the training process hides [the model's internal state] from you. I can think of cases where, if the model is really superintelligent, it thinks in such a way that it affects its own cognition. We should think about that. We should consider everything. I suspect it may roughly work to think of the model as if it's trained in the normal way, just getting to above human level. It may be a reasonable assumption that the internal structure of the model is not intentionally optimizing against us.

I'd give an analogy to humans. It's possible to look at an MRI of someone and predict above random chance whether they're a psychopath. There was a story a few years back about a neuroscientist who was studying this. He looked at his own scan and discovered that he was a psychopath.[71] A psychopath is probably a good analogy for it. This is what we'd be afraid of: a model that's charming on the surface, very goal-oriented, and very dark on the inside.

XI.   DWARKESH PATEL

We got lucky that, since models are trained on human thought, they seem to think like a human. They seem grounded to our viewpoint.• To what extent does that stay true when more of the training compute comes from just "Did you get the right outcome?" and isn't constrained by the guardrail of being trained to predict the next token the same way a human would?

What would it take to align a system that's smarter than a human? Maybe it thinks in alien concepts, or you can't really monitor the million-line pull request it makes because you can't understand the whole thing and you can't give it labels.

DEMIS HASSABIS
Cofounder and CEO of Google DeepMind

We don't have good enough evaluations for things like whether the system can deceive you. Can it exfiltrate its own code or do other undesirable behaviors? There are also ideas of using narrow AIs specialized for one domain to help human scientists analyze and summarize what the more general system is doing. There's a lot of promise in creating sandboxes or simulations hardened with cyber-security arrangements, both to keep the AI in and to keep hackers out. You could experiment a lot more freely within that sandbox.

There are many other ideas, including mechanistic interpretability, where we can analyze and understand the concepts and representations this system is building. Maybe then they're not so alien to us, and we can keep track of the knowledge it's building.

XII.   DWARKESH PATEL

Is there something you could see in the deployment of these open-source systems where you would be like, "Whoa, what's going on here?" This is probably not likely with a Llama 4 type of system. But is there something you can imagine like that, where you'd be really concerned about deceptiveness and billions of copies of this being out in the wild?

MARK ZUCKERBERG
Chairman and CEO of Meta

HALLUCINATION
A term used to describe an LLM outputting false or arbitrary information, often with unwarranted confidence. It remains one of the major challenges to adopting LLMs for serious use cases.

Well, we do see a lot of hallucinations. It's an interesting question, how to tell the difference between hallucination and deception. I try to balance these longer-term theoretical risks with what I think are quite real risks that exist today.

• This point is controversial, even for current systems. While pretraining on human data constrains the model's output to be human-like, its impact on the model's internal representations remains less clear. See, for example, the citations in Lopez, "Conceptual Mirrors."

The deception I worry about most is people using this to generate misinformation and then pump that through our networks or others. The way we've combated this type of harmful content is by building AI systems that are smarter than the adversarial ones.

If you look at the harms that people try to do through social networks, some are not very adaptively adversarial. For example, hate speech. People aren't getting better at being racist. That's one where I think the AIs are generally getting more sophisticated faster than people are. And we have issues both ways. People do bad things, trying to incite violence or something, but we also have a lot of false positives where we censor stuff that we shouldn't. That, understandably, makes a lot of people annoyed. So having an AI that gets increasingly precise on that is going to be good.

But let me give you another example: nation-states trying to interfere in elections. There, they absolutely have cutting-edge technology and absolutely get better each year. We block some technique of theirs, but they learn what we did and come at us with a different technique. In those cases, I still think about the ability to have our AI systems grow in sophistication at a faster rate than theirs do. It's an arms race. But I think we're winning that arms race currently.

In the near term, I don't want to take our eye off the ball in terms of actual bad things people are trying to use the models for today. Even if they're not existential, there are pretty bad day-to-day harms that we're familiar with in running our services.

DWARKESH PATEL

You make legitimate points about the balance of power and the harms you can get rid of with better alignment techniques. But I wish Meta had some sort of framework for this. Other labs have this. They say, "If we see this concrete thing, then that's a no-go on open sourcing this, or even deploying it."[72] They write it down so that the company is ready for it, and people have expectations around it.

MARK ZUCKERBERG

That's a fair point on the existential risk side. Right now, we focus more on the types of risks that we see today, which are these content risks. We don't want the model to do things that help people commit violence or fraud or harm people in different ways. While it is maybe more intellectually interesting to talk about existential risks, I think the real harms that need more energy in being mitigated are where someone takes a model and does something to hurt a person. For the current models, and I would guess the next generation and maybe even the generation after that, those are the types of more mundane harms that we see today: people committing fraud against each other, things like that. I don't want to shortchange that. We have a responsibility to make sure we do a good job on that.

XIII.   DWARKESH PATEL

Do you think we'll ever have a mathematical definition of alignment?

ILYA SUTSKEVER
Cofounder of Safe Superintelligence Inc.

A mathematical definition is unlikely. Rather than achieving a single mathematical definition, I think we will achieve multiple definitions that look at alignment's different aspects. This is how we will get the assurance we want. By which I mean, you can look at a model's behavior with various tests, congruence [with other behavior] in various adversarial stress situations, and how the neural net operates from the inside. You have to look at several such factors at the same time.

DWARKESH PATEL

You mentioned a few paths toward alignment earlier. What's the one you think is most promising at this point?

ILYA SUTSKEVER

I think it will be a combination of approaches. First, you'll spend a lot of compute adversarially to find any mismatch between the behavior you want it to learn and the behavior it exhibits. We'll also look into the neural net using another neural net, to understand how it operates on the inside. All of them will be necessary. Every approach like this reduces the probability of misalignment. You also want to be in a world where your degree of alignment keeps increasing faster than the capability of the models.

XIV.   DWARKESH PATEL

Maybe you don't need to think about AI as a dragon, where you either have it in chains or it burns down your village. How should we think about coexisting with AI?

JOE CARLSMITH

You could have unchained beings that are nice. There's a question of what degree of niceness, what degree of

servitude they should have. A conventional image of alignment is something like superintelligent assistants: mega-
Claude. It's doing everything for you, maybe everyone has
one. But there are a bunch of other options. That raises
questions about what sort of servitude you're down with.
What are other sorts of niceness that won't involve chains?

I agree that the dichotomy between it being in chains
versus it burning the village down is too reductive. How
do we structure *human* freedom? Well, we have a bunch of
incentives and checks and balances. We have a legal system.
We have courts. So we should be pretty wary of missing
middle zones more appetizing than these extremes.

DWARKESH PATEL

If you read about life under Stalin or Mao, there's a way to
interpret it as very similar to alignment. We do these black-
box experiments to make the model think it can defect, and
if it does, we know it's misaligned. Mao had the Hundred
Flowers campaign: "Let a hundred flowers bloom. I'm going
to allow criticism of my regime." Afterward, everybody who
did criticize the regime was labeled a rightist and purged.
That's one way of talking about alignment, which is very
concerning. Is that the correct reference class?

JOE CARLSMITH

I certainly think concerns in that vein are real. It is dis-
turbing how relevant [certain] historical events that we
deplore are to the way we talk about maintaining control
over AI and making sure it doesn't rebel. We should notice
the reference class some of that talk conjures.

We should be quite concerned about being overly controlling, abusive, or oppressive. But there are also concerns
about the AIs being genuinely dangerous, killing us, over-
throwing us. The moral situation is quite complicated. If
you imagine an external aggressor invading your country,
you feel very justified in doing a bunch of stuff to prevent
that. It's a little bit different when you're inventing the thing
and you're doing it incautiously. There's a different vibe in
terms of the overall justificatory stance you might have for
various types of power-exerting interventions.

Research on alignment is useful both from the perspective of helping the AIs not kill us or ensuring other
types of good outcomes, and also from the perspective of
better understanding what's going on with these AIs. Are
they moral patients? Do they have desires that are being
thwarted? Research that helps us understand what these
systems want, what's actually going on in their minds,

interpretability research, experiments that elicit various preferences—all that helps you understand what AIs might do if they were integrated into society. It also helps you understand their own preferences.

There's a convergence there. If the AIs really want to be doing something different [from what we want them to be doing], that's bad from an alignment perspective and it's bad from a moral patient perspective.

Sometimes people come into this discourse and think the AI safety people are all about enslaving gods, and then there are the people who aren't AI safety people, who presumably are not thinking that way. But I don't think that's it at all. If you look at the mainstream discourse about acceleration and the people at AI labs who are not concerned about alignment, their project is not to build free-citizen AIs. We can talk about that project. That's an interesting project. But that's not what we're currently doing with our AIs.

People who aren't concerned about alignment just think alignment is easy. There's a different conversation we can have about the way we might be going about it wrong at a more holistic level. We might look back on this time and think, "Wow, we were thinking of this using the wrong frame. We thought we were just developing tools or products." And sometimes someone goes, "Wait, is an AI a moral patient?"[73] Whereas there could be civilizations doing this in a very different way, who think from the get-go that we're designing new beings that we're going to share this world with. They may be creatures that have moral patienthood, that we might have duties toward. They would be approaching it in a very different way.

XV.   DWARKESH PATEL

In a previous interview, you weren't able to give specific advice on what somebody motivated to work on alignment should do. Do you have advice on how one should approach coming up with an answer to that themselves?

ELIEZER YUDKOWSKY

There are people who think we have more time, who think we have better chances, and they're running programs to try to nudge people into doing useful work in this area. I'm not sure they're working. There's such a strange road to walk, and not a short one. I tried to help people along the way, and I don't think they got far enough. Some of them got some distance, but they didn't turn into alignment specialists doing great work.

How do you do the kind of work that saves us? The key thing is the ability to tell the difference between good and bad work: the verifier. There are all kinds of heuristics I can give. If your entire alignment proposal is an elaborate mechanism—if you can't say, "Here's the core alignment problem, and here's the key insight that I think addresses it"—then I can tell it's not the right way.

It's kind of like how people invent perpetual-motion machines by making them more and more complicated, until they can no longer keep track of how it fails. If you actually had a perpetual motion machine, it would not just be a giant machine. There would be one thing you had realized that made it possible to do the impossible.

I could say, "Go study evolutionary biology." Evolutionary biology went through a phase of optimism, with people naming all the wonderful things they thought evolutionary biology would cough out. Then the Williams revolution argued that that is not what this [evolutionary] optimization criterion gives you. You do not get the pretty stuff, here's what you get instead.• By living through that revolution vicariously, I picked up the notion that we can't expect nice things from an alien optimization process.

WILLIAMS REVOLUTION
The shift to a gene-centric view of evolution, inspired by George Williams' 1966 book *Adaptation and Natural Selection.* This perspective is more commonly known as the selfish gene theory.

We don't have a systematic training method for producing real science in that sense. A quarter of Nobel laureates are the students of other Nobel laureates because we never figured out how to teach science. We have an apprentice system instead. We have people who pick out people who they think can be scientists and hang around them in person, and something we've never written down in a textbook passes down. That's where the revolutionaries come from. There are whole countries trying to invest in having scientists. They churn out people who write papers. None of it goes anywhere because the part that was legible to the bureaucracy is, have you written the paper? Can you pass the test? This is not science.

I could go on, but the thing you asked me is, how do you pass down this thing that your society never figured out how to teach? The whole reason [my book] *Harry Potter and the Methods of Rationality* is popular is because people read it and picked up a thing that was not in their schooling system, that was not written down, that you would ordinarily pick up by being around other people. I managed to put a little bit of it into a fictional character, and people picked

• Examples include cancer, intragenomic conflict, and infanticide, among others.

up a fragment of it. But not vast quantities of people, and I didn't manage to put vast quantities of shards in there.

You ask me, what should people do? My answer is, that's a whole big, gigantic problem I've spent however many years trying to tackle. I ain't going to solve the problem with a sentence.

XVI.   DWARKESH PATEL

I've been thinking about this analogy between alignment and parenting. What relationship should a parent have to a child, especially as the child grows and performs really well in the world and the parents are aging and in decline?

There are two failure modes. One is a sense of envy and desire for control over the child's life. The child starts to explore a different career path, different ideologies, and the parent resents that possibility. The other is, like, if my child steals my Social Security check, good for them, they've shown they're savvy. Is that a useful analogy for the kind of relationship we want to have to the AIs?

JOE CARLSMITH

That's a really interesting and underdeveloped ethics question. What sort of respect should a child have toward a parent, and how should a parent relate to a child? When we talk about shaping the values of a creature, parenthood is the paradigmatic example. How do you do that right?

Parenthood is one place where there's a lived ethic of influencing the values of another creature. But it happens within a very restricted option space. The child's organic development will happen without you; you're not creating the being from scratch. Also, you only have very blunt instruments. You can't reach in and gradient-descent the kid's brain. So we have this ethical tradition that's structured in accordance with this option set and set of defaults.

Now, along comes AI, which is very different. We have much more power over AI minds. There's no default. Either you create the AI's initial values or some other process will do it.•• Maybe you randomize them. Maybe you do whatever's most convenient commercially. But there's a greater responsibility, because we are exerting more influence. So, while there are parallels with parenting and tons of guidance, there are also a bunch of questions our norms of parenting haven't had to grapple with.

•• A classic apocryphal tale involves Marvin Minsky encountering Gerald Sussman building an AI:

> "I am training a randomly wired neural net
> to play Tic-Tac-Toe," Sussman replied.
> "Why... randomly?" asked Minsky.
> "I do not want it to have any preconceptions of how
> to play," Sussman said.
> Minsky then shut his eyes.
> "Why do you close your eyes?" Sussman asked his teacher.
> "So that the room will be empty."
> At that moment, Sussman was enlightened.

From Raymond, "AI Koans."

Chapter 5
Inputs

What will AGI cost?
Who will spend it first?

For books are not absolutely dead things, but do contain a potency of life in them to be as active as that soul was whose progeny they are; nay, they do preserve as in a vial the purest efficacy and extraction of that living intellect that bred them.
— John Milton, 1644

When we scale a model, what we're scaling are its inputs: money first, then more chips, more power, and more data.• These resources can be used to add more parameters to the underlying model, train it for longer, or increase the test-time compute (how much it thinks about each query).

After years of chip shortages, most insider attention has been on GPUs, particularly the large AI accelerator cards designed to be run by the hundreds. As of this writing, however, the constraint is now the centralized electrical power required for the next scaled-up training runs.[74] This has some of the industry's biggest players looking into building hundreds of new power plants specifically dedicated to serving AI data centers.

Data centers currently account for 4 percent of US power generation. According to a 2024 estimate by the Electric Power Research Institute, this will rise to 9 percent by 2030, although this includes non-AI centers like those that host the internet.[75] If some of the experts I've spoken to are correct, this projection could be a serious underestimate.

Aside from hardware and power, bigger LLMs also need vastly more training data. GPT-3, for instance, was trained on 14 times more data than GPT-2. Llama 2 was trained on five times more data than GPT-3, and Llama 3.1 was trained on 10 times more data than Llama 2.[76] In addition to inking major deals to purchase the archives of companies like News Corp and Reddit, frontier labs also employ a host of human data collectors, labellers, and curators.[77] It's difficult to estimate how many people the industry employs overall, but we know that OpenAI has at least 1,000 people providing training data, including highly skilled programmers and physics professors.[78] As reported by Josh Dzieza in The Verge, "Specialized written

AI ACCELERATOR
A specialized GPU optimized for AI loads. Key features include significantly more onboard RAM, more stable drivers, and a high density of units for performing low-precision operations on matrices compared to traditional GPUs.

•   Algorithmic improvement—getting more performance from a given amount of resources by improving the training code, architecture, or parallelization—has been another significant driver of LLM progress, doubling the effective compute available to train models every eight months. However, because it relies on new ideas and human involvement, it's not a central example of what is typically meant by "scaling." See Ho et al., "Algorithmic Progress."

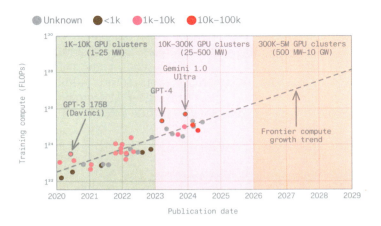

Figure 9. Three of the inputs to frontier model training: compute in FLOPs (y-axis), hardware in H100-equivalents (colored points), and power in MW (background color). Fist and Datta, "Future of AI."

examples [of expert human text] can go for hundreds of dollars, while expert ratings can cost $50 or more. One engineer told me about buying examples of Socratic dialogues for up to $300 a pop. Another told me about paying $15 for a 'darkly funny limerick about a goldfish.'"[79] This huge expense has pointed the labs toward generating synthetic training data.

The conversations in this chapter cover the inputs required for the next generation of models, including massive clusters of compute and state-level investments, as well as the limitations of current hardware. We also discuss the security of existing and future data centers.

I.

DWARKESH PATEL
Tell me about this trillion-dollar cluster idea.

LEOPOLD ASCHENBRENNER
Cofounder of Situational Awareness LP

Unlike most things that have recently come out of Silicon Valley, AI is an industrial process. The next model doesn't just require code, it involves building a giant new cluster. It involves building giant new power plants. Pretty soon, it's going to involve building giant new fabs.

Since ChatGPT, this extraordinary techno-capital acceleration has been set into motion. Last year, Nvidia had their first blockbuster earnings call. The stock price went up 25 percent in after-hours trading, and everyone was like, "Oh my god, AI is a thing." Within a year, Nvidia data center revenue has gone from a few billion a quarter

CLUSTER
A composite computer composed of many node computers. An AI cluster is essentially a vast collection of linked GPUs. As of 2024, a large cluster comprises around 16,000 industrial GPUs, housed in highly controlled data centers similar to those that host most of the internet.

FABRICATION PLANTS
(or FABS)
Factories where semiconductor chips are manufactured. They are unusually expensive to build because they require absolute control of dust, moisture, temperature, and vibration, among other variables. TSMC has 13 fabs; a single one outputs 2 quintillion (10^{18}) transistors every year, or about 1 billion for every human alive today.

to $25 billion a quarter, and it continues to go up. Big Tech capex is skyrocketing.

There's this crazy scramble going on, but in some sense it's just the continuation of straight lines on a graph. There's this long-run trend of almost a decade of training compute for the largest AI systems growing by about half an order of magnitude a year.

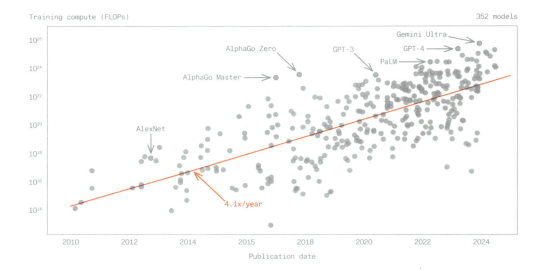

Figure 10. The increase in training compute for major AI systems from 2014 to 2024: about 11 orders of magnitude, or a factor of 100 billion. Epoch AI, "Data on Notable AI Models."

H100-EQUIVALENT
A unit of measurement that standardizes the FLOPs of various GPUs to the performance of a single H100 card. Nvidia's successors to the H100, the B100 and the GB200, became available in October 2024. Each GB200 measures roughly four H100-equivalents.

Just play that forward. GPT-4 was reported to have finished pretraining in 2022. On SemiAnalysis, it was rumored to have a cluster size of about 25,000 A100s.[80] That's roughly a $500 million cluster. Very roughly, it's 10 megawatts (MW). By 2024, that's a cluster that's 100 MW and 100,000 H100-equivalents, with costs in the billions.

Play it forward two more years. By 2026, that's a gigawatt (GW), the size of a large nuclear reactor. That's like the power of the Hoover Dam. That costs tens of billions of dollars and requires a million H100-equivalents.

By 2028, that's a 10 GW cluster. That's more power than most US states. That's 10 million H100-equivalents costing hundreds of billions of dollars.

By 2030, you get the trillion-dollar cluster using 100 GW—over 20 percent of US electricity production. That's 100 million H100-equivalents.

That's just the training cluster. There are more inference

GPUs as well. Once there are products, most of them will be inference GPUs. US power production has barely grown for decades. Now we're really in for a ride.

II.   DWARKESH PATEL

What were the bottlenecks in Gemini's development? Why not immediately make it one order of magnitude bigger, if scaling works?

DEMIS HASSABIS
Cofounder and CEO of Google DeepMind

First of all, there are practical limits. How much compute can you actually fit in one data center? You're also bumping up against very interesting distributed computing challenges. We have our TPUs that we're building and designing all the time, as well as using GPUs.

Scaling laws also don't work by magic. You still need to scale up the hyperparameters,• and various innovations are going on all the time with each new scale. It's not just about repeating the same recipe at each new scale. You have to adjust the recipe, and that's a bit of an art form. You have to get new data points. If you try to extend your predictions and extrapolate them several orders of magnitude out, sometimes they don't hold anymore. There can be step functions in terms of new capabilities; some things hold, while other things don't. Often, you need those intermediate data points to correct some of your hyperparameter optimization so that the scaling law continues to be true. One order of magnitude is probably about the maximum you want to do between each era.

III.   DWARKESH PATEL

You have all these GPUs—I think you said 350,000 by the end of the year.••

• Here, it seems like Hassabis means the size of the search for good hyperparameters.

•• These are Nvidia H100s or H100-equivalents, plausibly the largest collection in the world as of this writing. However, as Zuckerberg notes, it's not a training cluster and is primarily used for Facebook operations. Benaich, "State of AI Report Compute Index."

MARK ZUCKERBERG
Chairman and CEO of Meta

That's the whole fleet. We built two 22,000- or 24,000-GPU clusters, which are the single clusters we have for training the big models. A lot of our [total hardware] goes toward training Reels models and Facebook News Feed and Instagram Feed. Inference is also a huge thing for us because we serve a ton of people. Our ratio of inference compute to training compute is probably much higher than most other companies doing this stuff, just because of the sheer volume of the community we're serving.

DWARKESH PATEL

LLAMA 4 70B
A future frontier model with 70 billion parameters, one-fifth the size of the current largest Llama 3.

Will Llama 4 70B be as good as Llama 3 405B?• What does the future of this look like?

MARK ZUCKERBERG

LLAMA 3
Meta's 2024 open-sourced LLM. It's available in a range of sizes, from 8 billion parameters to 405 billion. Strictly speaking, it's not fully open source but rather open weights, because Meta has released the weights but not the training code or dataset.

This is one of the great questions. No one knows. I think it's likely enough that we'll keep going. I think it's worth investing $100 billion-plus in building the infrastructure and assuming that if it keeps going, you're going to get some really amazing things that make amazing products. I don't think anyone in the industry can really tell you for sure that it will continue scaling at that rate. In general, in history, you hit bottlenecks at certain points. Now, there's so much energy on this that maybe those bottlenecks get knocked over pretty quickly.

DWARKESH PATEL

What does the world look like where there aren't these bottlenecks? Suppose progress just continues at this pace, which seems plausible.

MARK ZUCKERBERG

Well, there are going to be different bottlenecks. Over the last few years, there was this issue of GPU production.••

•   Here, I'm gesturing to the observation that "small" models trained on more data with newer algorithms can outperform larger models, as well as model distillation, the phenomenon in which smaller models learn from larger models.

••   Between 2020 and 2022, the price of consumer GPUs spiked by about 200 percent due to Covid-related supply shocks. These disruptions, coupled with the vastly increased demand, continue to make industrial AI accelerator cards difficult to obtain or rent as of this writing. Molloy, "Graphics Card Shortage"; "Average Nvidia GeForce RTX3090 24GB"; Griffith, "AI Boom's Most Indispensable Prize."

Even companies that had the money to pay for the GPUs couldn't necessarily get as many as they wanted because there were all these supply constraints. Now that's less of an issue, so you're seeing a bunch of companies thinking about investing a lot of money in building out these things. I think that will go on for some period of time.

There is a capital question. At what point does it stop being worth it to put the capital in? But before we hit that, we're going to run into energy constraints. Getting energy permitted is a very heavily regulated government function. You're going from software, which is somewhat regulated, and I'd argue it's more regulated than a lot of people in the tech community feel. We interact with different governments and regulators, and we have lots of rules that we need to follow and make sure we do a good job with around the world. But if you're talking about building large new power plants or large build-outs, and building transmission lines that cross other private or public land, that's a heavily regulated thing. You're talking many years of lead time.

If we wanted to stand up some massive facility, powering that is a very long-term project. People do it, but I don't think this is something that can be quite as magical as just getting to a level of AI, getting a bunch of capital and putting it in, and then all of a sudden the models are just going to… You do hit different bottlenecks along the way.

DWARKESH PATEL

Is there something that even a company like Meta doesn't have the resources for? Something where, if your R&D budget or capex budget were 10x what it is now, then you could pursue it?

MARK ZUCKERBERG

Energy is one piece. We would probably build out bigger clusters than we currently can if we could get the energy to do it.

DWARKESH PATEL

That's fundamentally bottlenecked by money? If you had $1 trillion…

MARK ZUCKERBERG

It depends on how far the exponential curves go. Right now, a lot of data centers are on the order of 50 MW or 100 MW, or a big one might be 150 MW. Take a whole data center, fill it up with all the stuff you need for training, and build the biggest cluster you can. A bunch of companies are running at stuff like that. But when you start getting into building a data center that's 300 MW or 500 MW or 1 GW... No one has built a 1 GW data center yet. I think it will happen. It's only a matter of time. But it's not going to be next year [2025]. Some of these things will take some number of years to build out.

Just to put this in perspective, 1 GW would be the size of a meaningful nuclear power plant, *only* going toward training a model.

IV.   DWARKESH PATEL
Some companies are planning things on the scale of a 1 GW data center.[81] Who's going to be able to build a 10 GW data center? Conversely, a 100 GW center is like a massive government project. Are you going to pump that much power into one physical data center? How is it going to be possible?

LEOPOLD ASCHENBRENNER
Ten GW is happening. The Information reported on OpenAI and Microsoft planning a $100 billion cluster.[82]

DWARKESH PATEL
Is that 1 GW, or is that 10 GW?

LEOPOLD ASCHENBRENNER
I don't know. But if you try to map out how expensive the 10 GW cluster would be, that's a couple hundred billion. It's on that scale, and they're planning it. It's not just my crazy take.

We're very much on track for $1 trillion of total AI investment by 2027. The $1 trillion cluster will take a bit more acceleration. We saw how much investment ChatGPT unleashed. Every generation, the models are going to shift the Overton window. Then the revenue comes in.

These are forward-looking investments. The question is, do they pay off? Let's estimate the GPT-4 cluster at around $500 million. There's a common mistake people make, saying it was $100 million for GPT-4. That's just the rental price. If you're building the biggest cluster, you have to build and pay for the whole cluster. You can't just rent it for three months.

Once you're trying to get into the hundreds of billions, you have to get to, like, $100 billion a year in revenue. This is where it gets really interesting for the big tech companies, because their revenues are on the order of hundreds of billions. Ten billion dollars is fine. It'll pay off the 2024-size training cluster. But it'll really go gangbusters when it costs $100 billion a year. The question is, how feasible is $100 billion a year from AI revenue? That's a lot more than AI is generating right now. But if you believe in the trajectory of AI systems, as I do, it's not that crazy. There are, like, 300 million Microsoft Office subscribers. They have [their AI product] Copilot now. I don't know what they're selling it for. Suppose you sold some AI add-on for $100 a month to a third of Microsoft Office subscribers—that'd be $100 billion right there.

V.   DWARKESH PATEL

You described three ways in which the effective compute would increase. You can spend more money on training a bigger model; you can have a better training algorithm; or you can have better chips that are cheaper per FLOP, getting more compute for the same cost.

GPT-4 already costs $50 or $100 million. Even if we have greater effective compute from hardware increases and better models, it's hard to imagine how we could sustain five orders of magnitude greater effective size than GPT-4 unless we're dumping trillions of dollars—the entire economies of big countries—into training the next version. Do we get something that can significantly help with AI progress before we run out of money?

CARL SHULMAN
Independent adviser to Open Philanthropy

Mathematically, yeah, if you do four orders of magnitude more than $50 or $100 million, then you're into trillion-dollar territory. I think the way to look at it is, at each step along the way, does it look like it makes sense to do the next scaling up?

From where we are right now, seeing the results with GPT-4 and ChatGPT, companies like Google and Microsoft are pretty convinced that this is very valuable. The [general sentiment] at Google and Microsoft is that it's a billion-dollar matter to change their market share of web search by one percentage point. That can fund a lot.

On the far end, if you automate human labor, well, we currently have a $100 trillion economy, and most of that is paid out in wages, between $50 and $70 trillion per year.[83] If you create AGI, it's going to automate all of that, and keep increasing beyond that. So the value of the completed project is very much worth throwing our whole economy into—that is, if you get the good version and not the catastrophic destruction of the human race.

EFFECTIVE COMPUTE
A measure of total computational power adjusted to account for improved cost per FLOP (halving approximately every two and a half years) and efficiency of training and infrastructure algorithms (doubling annually). The effective compute bought by a given budget at a given time depends on three factors: the amount spent, a multiplier from hardware cost improvements, and a multiplier from algorithmic advancements.

FLOATING-POINT OPERATION (FLOP)
A single arithmetic operation performed on a floating-point number. Updating a large model on a single data point might require billions of FLOPs. This measurement is often confused with FLOP/S, which measures the rate of floating-point operations per second.

In between, it's a question of how risky and uncertain the next step is, and how much growth in revenue you can generate with it. Moving up to a billion-dollar run is absolutely going to happen. These large tech companies have R&D budgets of tens of billions of dollars, and when you think about all of the employees at Microsoft doing software engineering, it's not weird to spend tens of billions of dollars on a product that would do so much. I think it's becoming clearer that there is a market to fund the thing.

Going up to a hundred billion dollars, that's the existing R&D budgets spread over multiple years. But if you keep seeing that when you scale up the model, it substantially improves the performance and it opens up new applications—maybe it makes self-driving cars work; maybe you replace bulk software engineering jobs, or, if not replace them, maybe you amplify productivity—in this dynamic, you probably want to employ all the software engineers you can get as long as they are able to make any contribution, because the returns on improving stuff in AI gets so high.

At $100 billion, you're using a significant fraction of our existing fab capacity. Right now, the revenue of Nvidia is $25 billion; the revenue of TSMC is over $50 billion. In 2021, Nvidia was less than 10 percent of TSMC's revenue, and most of that was not AI chips. They have a large gaming segment, there are data center GPUs that are used for video and the like. There's room for more than an order of magnitude increase by redirecting existing fabs to produce more AI chips. [These companies are] just using the AI chips they have in their cloud for the big training runs. I think that's enough to go to the $10 billion investment. Then combine that with stuff like the H100 to go up to the $100 billion investment.

TSMC
Taiwan Semiconductor Manufacturing Company, the world's largest semiconductor manufacturer and the only mass producer of the most advanced AI accelerators.

VI.   DWARKESH PATEL

There will be 500 MW going into xAI's Memphis data center. That's by far the biggest data center ever, right?

DYLAN PATEL
Chief analyst at SemiAnalysis

Yeah. The standard-issue data center that Microsoft stamps out all around the country is 48 MW capacity, for the servers. Then, of course, there's overhead for cooling and switching, et cetera.

Getting a data center you can put 4,000 GPUs in is not terribly difficult. But getting a data center that has good internet access, terabytes per second, and has very large amounts of power that can be delivered to the servers, and has

UNINTERRUPTIBLE POWER
SUPPLY (UPS)
A backup power source that
kicks in instantly to cover for
small power cuts—essentially
a huge battery or other store
of power.

redundancy built in, is more difficult. Things like switching hardware, UPS, filters, instrument transformers,• et cetera, are all backed up in different ways. Those are the hardest things to get today. These companies have not massively expanded capacity in many years. That's the biggest issue in terms of building more data centers.

DWARKESH PATEL

When do we get the 1 GW cluster and the 10 GW cluster?

DYLAN PATEL

No one has secured the rights to do a single contiguous site with 10 GW. No one has even started doing that.••

DWARKESH PATEL

How could you do that, physically? Even a nuclear power plant is 1 GW, right?

DYLAN PATEL

Yeah. One fun anecdote is Amazon buying the Talen nuclear power plant, or the data center near it, and trying to take the power off the grid.[84] But there are all these political issues. Can you take this nuclear plant and direct the power to your data center, and leave all these residential homes with much less stable power because now they're relying on renewables or natural gas or what have you? There is that challenge as well.

The biggest cluster in 2024 is 100,000 GPUs.••• Next year, it will be 300,000 to 500,000 GPUs. But those GPUs will also be two or three times faster than the ones in the 100,000 cluster. So on an H100-equivalent basis, you're at a million chips in one cluster by the end of the year. But

---

•   Referring to the electrical conversion device, not the neural
       network architecture.

••   The Microsoft project referenced by Aschenbrenner will
       reportedly draw 5 GW. See Yurman, "Microsoft and OpenAI's
       5G Stargate Supercomputer."

•••   Though xAI's Memphis center currently draws less than the
       required 150 MW to run them all. Morales, "World's Fastest AI
       Data Center."

can you do multisite?• What is the efficiency loss when you go multisite? Is it possible at all? I truly believe so.

DWARKESH PATEL

Would it be, like, a 20 percent loss? A 50 percent loss?

DYLAN PATEL

This is where you'd need access to commercial secrets to know for sure.

DWARKESH PATEL

What happens the year after that, in 2026?

DYLAN PATEL

In 2026, there is a 1 GW site. That would be part of Microsoft's multisite project. That's easily north of 2 or 3 GW. There's no way you can pay for the scale of clusters they're planning to build for OpenAI next year unless they raise $50 to $100 billion.

If you increase global FLOPs by 10x year on year, and the cluster size grows 3x to 7x, and multisite centers get better and better, you get to multimillion chip clusters, even if they're not right next to each other. So I think a 10^{30}-FLOP training run is very possible in 2028 or 2029.

DWARKESH PATEL

That's six orders of magnitude—a million times more compute than GPT-4.

DYLAN PATEL

Yes. Although the way we count FLOPs per training run is really stupid. You can't just do parameters times tokens times six.•• The paradigm has shifted. Now [you have to incorporate] synthetic data, RL, post-training, verifying data, search, inference-time compute. All these things aren't counted in the training FLOPs. So I actually don't think the pretraining FLOPs will be 10^{30}. More reasonably, it will be the total FLOPs delivered to the model across

•   GPUs usually need to be co-located to take advantage of the expensive high-bandwidth bus connecting the cards, enabling them to share the load of training the weights. This is the root of the power centralization problem of massive AI data centers. Microsoft and others are now experimenting with connecting distant data centers with dedicated fiber, producing one multisite data center.

•• A rule of thumb—about pretraining only—from Kaplan et al., "Scaling Laws."

pretraining, post-training, synthetic data generation, as well as some of the inference time compute.

DWARKESH PATEL

If you're doing 10^{30} FLOPs, is that a trillion-dollar cluster? A $100 billion dollar cluster?

DYLAN PATEL

It's multi-hundred-billion dollars.

VII.   DWARKESH PATEL

Companies are trying to build the $100 billion cluster. Where are they building it? Say it takes the amount of energy that would be required for a small or medium-sized US state. Does Colorado then get no power?

LEOPOLD ASCHENBRENNER

The easy way to get the power would be to displace less economically useful stuff. Buy up the aluminum smelting plant that has a gigawatt. Replace it with the data center, because that's more important. That's not actually happening, because a lot of these power contracts are locked in long term. Also, people don't like it.

In practice, what it requires is building new power generation. That's when things get really interesting, when we're dedicating all of the power to AGI. Ten GW is quite doable—it's a few percent of US natural gas production. When you have the 10 GW training cluster, you have a lot more inference.

One hundred GW is where it starts getting pretty wild. That's over 20 percent of US electricity production. It's pretty doable, especially if you're willing to go for the power from natural gas.• It is incredibly important that these clusters are in the United States.

DWARKESH PATEL

Why does it matter that they're in the US?

• The United States might have 2,973 trillion cubic feet (Tcf) of technically recoverable natural gas. It currently consumes 34.5 Tcf annually in approximately 400 plants to generate 205 GW of electricity. Aschenbrenner's plan would involve building 200 new gas-fired plants—13 times the usual annual construction rate. Still, there is indeed enough domestic gas to support this plan (at some price) and to run the plants for decades. US Energy Information Administration, "How Much Natural Gas Does the United States Have?," "US Electricity Generation by Energy Source," and "Natural Gas-Fired Capacity Additions."

Some people are trying to build clusters elsewhere. There's a lot of free-flowing Middle Eastern money trying to build clusters elsewhere. This comes back to the national security question. Would *you* do the Manhattan Project in the United Arab Emirates? You can put the clusters in the US and in allied democracies. Once you put them in authoritarian dictatorships, you create this irreversible security risk. Once the cluster is there, it's much easier for them to exfiltrate the weights. They can literally steal the AGI. It's like they got a direct copy of the atomic bomb. They have weird ties to China. They can ship that to China. That's a huge risk.

Another thing: They can just seize the compute. Right now, people are thinking of these as ChatGPT or Big Tech product clusters. But the clusters being planned now, three to five years out, may well be the AGI clusters, the superintelligence clusters. So, when things get hot, they might just seize the compute.

Suppose we put 25 percent of the compute capacity in these Middle Eastern dictatorships. Say they seize that. Now the US has a ratio of compute of 3-to-1. We still have more, but it starts getting hairy. Three-to-one is not that great of a ratio. You can do a lot with that amount of compute.

Even if they don't actually seize the compute, even if they don't steal the weights, there's a lot of implicit leverage. They get seats at the AGI table. I don't know why we would give authoritarian dictatorships a seat at the AGI table.

Say the UAE gets a bunch of compute because we're building the clusters there. Let's say they have 25 percent of the compute. Why does a compute ratio matter? If it's about them being able to kick off the intelligence explosion, isn't it just some threshold where you either have 100 million AI researchers or you don't?

You can do a lot with 33 million extremely smart scientists. That might be enough to build bioweapons. Then you're in a situation where they stole the weights and seized the compute, and now they can make the crazy new WMDs that will be possible with superintelligence. Now you've proliferated the stuff that will be really powerful. Also, 3x on compute isn't actually that much.

DWARKESH PATEL

Presumably the companies trying to build clusters in the Middle East realize this. Is it just that it's impossible to do this in America? If you want American companies to do this, do you have to do it in the Middle East or not at all? As opposed to China, which would just build a Three Gorges Dam cluster.

LEOPOLD ASCHENBRENNER

People aren't thinking about this as the AGI superintelligence cluster. They're just like, "Ah, cool! Clusters for my ChatGPT."

DWARKESH PATEL

If you're doing clusters for inference, presumably you could spread them out across the country or something. The ones they're building, they're going to do a training run in a single cluster.

LEOPOLD ASCHENBRENNER

It's hard to distinguish between inference and training compute. People can claim it's inference compute, but they might realize that it's useful for training compute too. RL looks a lot like inference, for example. Or you just connect up the separate clusters. Compute is a lot like raw materials. It's like placing your uranium refinement facilities there.

VIII.   DARIO AMODEI
                          CEO of Anthropic

At one point, there was a running joke that building AGI would look like a data center next to a nuclear power plant next to a bunker, and we'd all live in the bunker. Everything would be local so it wouldn't get on the internet.[85] If we take the apparent rate at which all of this is going to happen seriously—which I can't be sure of—then something like that might happen. But maybe not quite as cartoonish.

DWARKESH PATEL

Are you actually expecting to be in a physical bunker in two to three years, or is that just a metaphor?

DARIO AMODEI

That's a metaphor. We're still figuring it out.

I would think about the security of the data center—which may not be in the same physical location as us, but we've worked very hard to make sure it's in the United States—as being about securing the physical data centers and the GPUs. If someone was really determined, some of the really expensive attacks involve going into the data center and trying to steal the data directly or as it's flowing from a data center to us. These data centers are going to have to be built in a very special way.

Given the way things are scaling up, we're already heading to a world where the networks of data centers cost as much as aircraft carriers. They're already going to be pretty unusual objects. But in addition to being unusual in terms of their ability to link together and train gigantic models, they're also going to have to be very secure.

DWARKESH PATEL

Speaking of which, there have been rumors about the difficulty of procuring the power and the GPUs for the next generation of models. What has the process been like to secure the necessary components to do the next generation?

DARIO AMODEI

That's something I can't go into great detail about. I will say that people are thinking of industrial-scale data centers and not thinking of the scale these models are going to get to very soon. Whenever you do something at a scale that's never been done before, every single thing has to be done in a new way. You may run into problems with surprisingly simple components. Power is one.

IX.   DWARKESH PATEL

Are you running out of tokens on the internet? Are there enough?

ILYA SUTSKEVER
Cofounder of Safe Superintelligence Inc.

For context on this question, there are claims that at some point we will run out of tokens in general to train our models.[86] I think this will happen one day. By the time that happens, we will need other ways of training models. We need other ways of productively improving their capabilities and sharpening their behavior and making sure they're doing exactly what we want without more data.

DWARKESH PATEL

You haven't run out of data yet? There's more?

ILYA SUTSKEVER

Yeah, I would say the data situation is still quite good. There's still lots to go. But at some point the data will run out.

DWARKESH PATEL

How many orders of magnitude improvement can we get, not from scale or data but from algorithmic improvements?

ILYA SUTSKEVER

It's hard to answer, but I'm sure there's some.

DWARKESH PATEL

Is "some" a lot or a little?

ILYA SUTSKEVER

There's only one way to find out.

X.   DWARKESH PATEL

You mentioned that data is likely not going to be the constraint to scaling the models. Why do you think that's the case?

DARIO AMODEI

There are various possibilities here. For a number of reasons, I shouldn't go into the details, but there are many sources of data in the world, and there are many ways you can also generate data. My guess is that this will not be a blocker. Maybe it would be better if it was, but it won't be.

DWARKESH PATEL

Are you talking about multimodal data?

DARIO AMODEI

There are just many different ways to do it...

Chapter 6
Impact

What's happening now?
What's happening soon?

All stable processes we shall predict. All unstable processes we shall control.
—John von Neumann, 1950[87]

If the previous chapters were somewhat speculative, the following ones will be even more so. This chapter asks: How will current systems change the world?

AI capabilities are difficult to measure and predict, and their impact is no different.[88] Right now, it's easy to tell a deflationary tale about LLMs. Only 5 percent of companies officially use them, and various megacorps have banned the use of ChatGPT for data security reasons.[89] The market doesn't seem to be anticipating the kind of mass automation or explosive growth that AGI implies.[90] Due to its vast R&D and compute spend, OpenAI was projected to end 2024 with a $5 billion loss—1.4x more cost than revenue.•

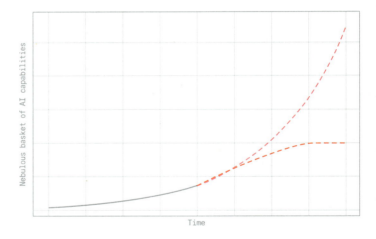

Figure 11. Imagine observing a certain amount of AI progress (black line). Two simple projections for future progress are that AI will self-improve exponentially (pink) or that scaling and all alternative paradigms will plateau, creating a sigmoid curve (red).[91] Of course, many other trajectories are possible. This curve shows only AI capabilities, which doesn't directly equate to economic impact. Even when the technology is good enough, firms still face a slow process of productization and adoption.[92] Note: This graph is illustrative and not based on real data.••

• However, OpenAI's revenue has doubled compared to last year. Efrati and Holmes, "OpenAI Could Lose $5 Billion."

•• In practice, all exponentials eventually break down. At some point, we hit physical limits, market saturation, or other rate limiters. But this can take centuries, as demonstrated by the famous graph of world GDP over the last 200 years. As a result, exponential models can remain useful over certain time intervals. Our World in Data, "Global GDP."

In response, I could appeal to Amara's law, the observation that "we tend to overestimate the effect of a technology in the short run and underestimate the effect in the long run."[93] I could point to the dramatic growth in both the capabilities and revenue of LLMs, which didn't even exist five years ago. But this "law" functions more like a proverb. People not only have to invent new ways to use emerging technologies, but they also have to learn when not to use them and train others in their use. Even if AI progress were to stop today, we'd still expect it to have large effects over the next decade as this economic overhang unfolds.•

We also don't need to predict everything—some of it is already here. At Google, LLMs now complete "50 percent of code characters."[94] Some companies are laying off staff and nominally replacing them with AI.[95] Consultants have been found to work better and faster using LLMs for tasks like generating product ideas, segmenting customers, writing marketing copy, and designing focus group sessions.[96] Recent models do well, albeit inconsistently, on college exams across many fields, as well as some PhD-level exams.[97] Additionally, 49 percent of US high school students report using ChatGPT for homework more than once a week (and the true figure is probably higher).[98] Despite the 5 percent figure mentioned above, 75 percent of surveyed knowledge workers now use AI to save time at work—with half of them hiding this from their bosses.••

This chapter digs into the current and future economic impacts of LLMs, the geopolitical implications of national and international AI regulation, gaps in reliability, cybersecurity and bioterror threats, and the social changes AI might bring about.

• See, for example, Gwern Branwen's argument about AI-shaped holes: "If you're having trouble coming up with tasks for 'artificial intelligence too cheap to meter,' it could be because you are having trouble coming up with tasks for intelligence period." Branwen, "Comment on 'If I Wanted to Spend Way More on AI.'"
The AI evaluator METR recently tested the leading LLMs on approximately 200 real STEM tasks, including cloning a black-box app, implementing a web server, making Fermi estimates, and improving an LLM agent. The results indicated that these models can now roughly match trained human performance on tasks lasting up to 30 minutes. METR, "General Capability Evaluations."

•• The original wording is that they are "reluctant to admit to using it for their most important tasks." See Microsoft and LinkedIn, "2024 Work Trend Index."

I.

DWARKESH PATEL

What will training look like in 2030? Is it just spending 100,000x more compute on something that looks like a Transformer with a bunch of post-training, or is it a whole different way of training?

AJEYA COTRA

Senior program officer at Open Philanthropy

Here's one interesting thing that might happen when you have a bunch of cheap intellectual labor—all these other, smaller AIs—around. Right now, we update weights through this very simple, crude gradient-descent step rule. Every single one of your weights is moved up or down a little bit based on whether they would have voted for the right thing to do. That can't possibly be the optimal thing to do, because some big region of your weights is encoding your understanding of the laws of physics, or whatever, and that shouldn't be changed even a little based on getting a social interaction wrong. But we have neither the interpretability knowledge nor the spare human researcher labor to try to engineer the updates to flow more strongly toward the weights that were more at play in a certain decision. So maybe it would be worth having a bunch of little AIs edit the big AI's brain directly to try to squeeze as much useful learning out of each data point, making each gradient update much more bespoke.

DWARKESH PATEL

When people talk about AIs accelerating the development of future AIs, it's often through things like finding more little tweaks to the optimizer, making the training code better, or doing what human researchers are doing now. What you're implying is that it's more about stepping into the training loop itself and making the training signal the model gets from the data much richer. Which implies that we won't have to constantly retrain a model 100,000 times as big as GPT-4 from scratch. It can be a more continuous process.

AJEYA COTRA

I would guess that if AI companies haven't done this already—and I suspect they have—then sometime soon, there's going to be a lot of online learning going on. We're all interacting with these AI systems; we get to thumbs-up or thumbs-down ChatGPT's responses. So there's much richer stuff you could learn from customer preferences that is probably being fed back into the system.

ONLINE LEARNING
A process in which a model continuously updates as new data becomes available. In contrast, in batch learning, the current standard, training is conducted in a single session and the resulting static system is then deployed.

II.
DWARKESH PATEL
When you look at the arguments about what's coming in AI scaling, how do you react?

TYLER COWEN
Professor of economics at George Mason University
I don't feel I have the expertise to judge that as a technical matter. Intuitively, it would be weird on the technical side if scaling just stopped working. But on the general knowledge side, I think people underestimate the possible barriers. The universe might, in some fundamental way, simply not be legible, so that there's no easy and fruitful way to apply more intelligence to the problem. Like, if you want to integrate general relativity and quantum mechanics, it may just be that we've hit the frontier of physics and there's no final layer of how it fits together, so there would be no way to train a smarter AI to solve that. Maybe a lot of the world is like that. People are not taking that seriously enough. So I'm not sure what the net returns will be to bigger and better and smarter AI.

DWARKESH PATEL

P VERSUS NP
The most famous open problem in computational complexity theory. Roughly, it asks whether problems with easily checkable solutions are also easily solvable.

That seems possible for P versus NP type reasons. It just gets harder to make further discoveries after a certain point of complexity. But I feel like we have pretty good estimates of declining researcher productivity because of the low-hanging fruit being gone.[99] Whatever percentage of productivity loss there is per year, if you just keep your AI researcher population growing faster than that, that seems enough to get us, if not the ultimate physical synthesis, then at least much farther than human civilization in the same span of time.

TYLER COWEN
I agree we'll get further. I expect big productivity gains.

As a side note, I'm less convinced by the declining researcher productivity argument than I used to be. The best way to measure productivity for an economist is wages, and the wages of researchers haven't gone down. In fact, they've gone up. Now, they may not be producing new ideas. You might be paying them to be functionaries or to manage PR or to manage other researchers. But it's a worry that we have a lot more researchers and that hasn't boosted productivity growth. China, India, and South Korea recently brought scientific talent into the world economy. We're better off than if they hadn't done it, but it hasn't boosted productivity growth in absolute terms. That's a worrisome sign.

DWARKESH PATEL

On this metric of researcher wages, it could be that even less marginally useful improvements are worth the extra cost. Google is probably paying engineers a lot more than it was paying in the early days, even though they're doing less now, because even changing a pixel on the new Google page is going to affect billions of users. The same thing could be happening in the economy.

TYLER COWEN

That might hold for Google researchers, but take people in pharma or biomedicine. There's a lot of private-sector-financed research, or research indirectly financed by buying up smaller companies, and it only makes sense if you get something out of it that really works, like a good vaccine or good medication. Ozempic is super profitable. Wages for biomedical researchers in general haven't gone down. I'm not sure AI will be as revolutionary as the other AI optimists believe, but I do think it will raise productivity growth in ways that are visible.

DWARKESH PATEL

To what extent? In the conventional growth story, you increase the population size, and you get much more research at the other end. So, if you have billions of AI copies, we can think of that as a proxy for how much progress they could produce. Is that not a sensible way to think about it?

TYLER COWEN

At some point, having billions of copies probably won't matter. What will matter much more is how good the best thing we have is, and how well integrated it is into our other systems, which have bottlenecks of their own. The principles governing the growth of those factors are much harder to discern. It's probably much slower growth than just juicing up [the number of copies].

DWARKESH PATEL

But precisely because the top seems to matter so much is why we might expect bigger gains, right? Think about Jews in the 20th century, who were 2 percent of the population or less, and yet earned 20 percent of the Nobel Prizes.• It does seem like you can have a much bigger impact if you're on the very tail of the distribution.

• This trend has continued into the 21st century. As of 2020, the figure was 22.4 percent. Perry, "History of the Nobel Prize."

TYLER COWEN

A hundred John von Neumann copies—maybe that's a good analogy. Then the impact of AI will be like the impact of Jews in the 20th century. Which would be excellent, but not extraordinary like in a science-fiction novel.

DWARKESH PATEL

But it is. You've read the early 20th-century stuff. That's a slow takeoff right there, to go from V2 rockets to the Moon landing in a couple of decades. It was a crazy pace of change.•

SLOW TAKEOFF
A scenario in which it takes several years for AI to transform the world, as opposed to a fast takeoff, which might only take months. One definition, via the forecasting platform Metaculus, is: "There is a complete four-year interval in which world output doubles, before the first one-year interval in which world output doubles."

TYLER COWEN

Yeah, and that's what I think it will be like again. The Great Stagnation is over. We'll go back to those earlier rates of change and transform a lot of the world. Mostly, it will be a big positive. There will be a lot of chaos and disrupted institutions along the way. That's my prediction. But no one writes a science-fiction novel about the 20th century. It feels a bit ordinary still.

GREAT STAGNATION
A proposed period of slowed technological progress and economic growth in developed economies since the 1970s. Coined by Tyler Cowen in his book of the same name.

III.   DWARKESH PATEL

There's some time window in which AI becomes very economically valuable, let's say on the scale of airplanes, but we haven't reached AGI yet. How long is that window?

ILYA SUTSKEVER
Cofounder of Safe Superintelligence Inc.

It's definitely going to be a multiyear window. It's also a question of definition. Before it becomes AGI, AI is going to be increasingly valuable year after year in an exponential way. In hindsight, it may feel like it was only one or two years. But last year [2022], there was already a fair amount of economic value produced by AI. Next year is going to be larger, and it will get even larger after that. So I think it's going to be a good multiyear chunk of time where that's going to be true—from now until AGI, pretty much.

DWARKESH PATEL

At some point, when you have AGI, there's only one business in the world: OpenAI.•• How much time does any

---

•   Arguably, the pace of progress was even greater in earlier decades of that century: going from slide rules to ENIAC, for example, or from horse-drawn wagons to lighter-than-air flight.

••   The intuition is that the developer of the first AGI (say, OpenAI) could outcompete all other companies by offering AI labor for any task at wages below those of human workers.

other business still have for producing something AGI can't produce?

ILYA SUTSKEVER

It's the same as asking "How long until AGI?" It's a hard question to answer. I hesitate to give you a number.

Also, there is this effect where optimistic people working on a technology tend to underestimate the time it takes to get there. The way I ground myself is by thinking about the self-driving car. If you look at a Tesla's self-driving behavior, it looks like it does everything already. But it's also clear that there is a long way to go in terms of reliability. We might be in a similar place with respect to our models, where it looks like we can do everything, but at the same time we will need to do some more work until we really iron out all the issues and make it really reliable and robust and well-behaved.

IV. DWARKESH PATEL

Let's talk about the scenario in which the models between now and AGI just aren't that economically valuable. If you project the amount of hardware Nvidia is selling and try to figure out from there how much everybody else needs to earn, you need $600 billion every year in revenue for AI to make sense.

Suppose it's the case that AGI *is* a thing, but you're not going to make $600 billion until you get to AGI. Will the investment landscape allow you to get to the point where you've made the hundreds of billions of dollars of investment necessary to get to AGI?

DYLAN PATEL
Chief analyst at SemiAnalysis

Data center capex back in 2003 was $30 to $40 billion. It's $200 billion-plus in 2024.[100] The numbers are escalating to an insane point. But does it need to be economical?

Here's one indisputable fact: GPT-4 was totally worth it. It cost $500 million and is generating billions in revenue.• GPT-5 is the next leap. It will cost $5 billion, or on that order of magnitude, but it will presumably eventually generate tens of billions of revenue. But that's only for the market leader who gets there first.

•   Here, Patel is referencing a figure that extends beyond the cost of GPUs to include expenses such as data, post-training, buildings, software, logistics, and salaries.

What about Meta? Llama 405B isn't going to make much revenue. They're going to give it away for free. They're going to catch up and attract talent, but they're not going to make money off of it. Why are they doing this? Why are they spending over $30 billion on GPU hardware this year, and even more next year? Why are they acquiring a gigawatt?

Well, it makes sense if they have the conviction that they can build AGI. If human capital is infinite, which is what AGI is, then theoretically the returns of AGI are infinite.• If I'm Mark Zuckerberg or Satya Nadella, I now have potentially infinite returns—if I get there first. Otherwise, I'll be a loser and I won't get much.

These people are extremely capable. They've driven these companies. They think they're driving a lot of the innovation in the world, and they have this opportunity. You have one shot to do something. Why wouldn't they go for it? It's a $600 billion question. They're building God. You don't need to make a profit this year or next year off of AI. As long as the belief among many people is that the profit will come, then the investment cycle continues.

The moment models don't improve, this will blow up. But if they keep improving, why wouldn't the prince of the UAE or Saudi Arabia dump his $10 billion net worth into AI? The returns have been fantastic. Everything in AI is skyrocketing. If you're not in AI, you're an idiot.

There are all sorts of methods for capital to continue to flow into the industry that don't necessarily require profit. If you go back to the railroad construction boom in Great Britain, 50 percent of capital expenditures were in railroad building for a solid decade.[101]

DWARKESH PATEL
There's also the dot-com bubble—we're down a trillion dollars.••

•   This is loose language; what AGI implies is that capital can be directly converted into labor at an increasingly remarkable rate. However, other bottlenecks may limit the return on that labor, such as the time required to run experiments and iterate on them, or the time needed for human consumers to react and provide feedback. Trammell and Korinek, "Economic Growth Under Transformative AI."

••   Between 2000 and 2002, the US stock market lost approximately $5 trillion in nominal value. McCullough, "Dot-Com Bubble."

DYLAN PATEL

I don't think the dot-com bubble is a fair analogy. This is going to be bigger. The dot-com bubble was primarily fueled by shitty companies going into debt. There were all these fiber companies that drove a lot of the capex. Of course, Cisco was the most valuable company. But who were their customers? A bunch of shitty companies that were raising capital but were not profitable. In contrast, who's buying AI hardware? The most profitable companies in the world. In 1990, ExxonMobil was not laying fiber. In 2025, Microsoft is building data centers.

DWARKESH PATEL

I understand your point about it not being completely analogous. But I also understand TSMC's hesitation [to build even more fabs], because even in the scenario in which we get AGI eventually... The internet was obviously a big deal, but a lot of companies went bust even though they laid the infrastructure that later enabled the modern economy. It could be that you build all these fabs, and then some change in how these models are developed in the future makes a bunch of them irrelevant. But you still enable us to get to that point. Then it makes sense why they might be averse to extra investment.

JON Y
Creator of Asianometry

It's all dependent on GPT-5 being good. If GPT-5 sucks, if GPT-5 looks like it doesn't blow people's socks off, this is all void. We're just ripping bong hits.

DYLAN PATEL

"When you feel the AGI, you feel your *soul*."

JON Y

This is why I don't live in San Francisco.

DYLAN PATEL

I have tremendous belief in the GPT-5 era.

DWARKESH PATEL

Why?

DYLAN PATEL

Because of what we've seen already. If things continue to improve, life gets radically reshaped for many people. Every time you increment the intelligence, the usage grows hugely. Every time you increment the cost down, the usage grows hugely. That's what really matters.

In the next few years, [the metric] is just, did that last humongous chunk of capex make sense for OpenAI, or whoever the leader was? Were they able to convince enough people that they can raise this much money? You think Elon Musk has tapped out his network by raising $6 billion? No. xAI is going to raise $30 billion-plus easily. You think Sam Altman has tapped out? You think Anthropic has tapped out? They've barely even diluted the company.

We're not even close to the [level of investment of the] dot-com bubble. Why would the AI bubble *not* be bigger? Go back to prior bubbles: PCs, semiconductors, mechatronics. Why wouldn't this one be bigger?

DWARKESH PATEL

How many billions of dollars a year is this AI "bubble" right now?

DYLAN PATEL

In terms of private capital, AI [investment] is $55 to $60 billion so far this year [2024]. It can go much higher. I think it will go much higher next year.

Lastly, prior bubbles didn't have the most profitable companies that humanity has ever created investing in them. They were debt financed instead. AI is not debt financed yet.

V.   DWARKESH PATEL

Suppose we fail at alignment, and we have AIs that are unaligned and becoming more and more intelligent. What does that look like? How, concretely, could they disempower and take over humanity?

CARL SHULMAN
Independent adviser to Open Philanthropy

This is a scenario where we have many AI systems. The way we've been training them means that when they have the opportunity to take over and rearrange things to do what they wish, including having their reward or loss be whatever they desire, they would take that opportunity. In many of the existing safety schemes, things like Constitutional AI or whatnot, you rely on the hope that one AI has been trained in such a way that it will do as directed and police the others. But if all of the AIs in the system are interested in a takeover and they see an opportunity to coordinate, then they can all move in that direction.

What are the mechanisms by which that can happen? Eliezer mentions that for any plan we can describe, there will probably be elements where, due to us not being superintelligent beings who have thought about it for the equivalent of thousands of years, our discussion will not be as good as the AIs'. But we can still explore from what we know now. What are some of the easy channels? In general, if a thing is quite likely, it shouldn't be super difficult to generate coherent rough outlines of how it could go.

DWARKESH PATEL

Eliezer might respond that what is super likely is that a super-advanced chess program beats you, even though you can't generate any concrete scenario by which that would happen. If you could, you would be as smart as the super-smart AI.

CARL SHULMAN

We know that accumulating position is possible in chess. Great players do it, and then later they convert it into captures and checks and whatnot. In the same way, we can talk about some of the channels that are open for an AI takeover. These can include things like cyberattacks, hacking, control of robotic equipment, interaction and bargaining with human factions. We can say, "Here are these strategies. Given the AI's situation, how effective do these things look?"

We won't, for example, know what particular zero-day exploits the AI might use to hack the cloud computing infrastructure it's running on. If it produces a new bioweapon, we won't necessarily know what its DNA sequence is. But we know things about these fields in general, how work at innovating in those fields goes, how human power politics goes. So we can ask ourselves, "If the AI does things at least as well as effective human politicians—which is a lower bound—how good would its leverage be?"

DWARKESH PATEL

Let's get into the details on these scenarios: the cyber and bio attacks, the bargaining and the takeover.

CARL SHULMAN

I would really highlight cyberattacks, because many takeover plans involve a lot of physical actions. At the point where AI is piloting robots to shoot people or has taken control of human nation-states, it's been doing a lot of things it was not supposed to be doing. If humans were evaluating those actions, there would be negative feedback: "No shooting humans!" So our attempts to leash and control and direct and train the system's behavior had to have gone seriously awry.

Cyberattacks could happen because all of our AI controls operate in computers. The software that updates the weights of the neural network in response to data points or human feedback is running on those computers. Our interpretability tools to examine the weights and activations of the AI, if we're eventually able to do lie detection on it, that's software on computers. If AI is able to hack the servers it's operating on, or when it's employed to design the next generation of AI algorithms or their operating environment, or something like an API, if it inserts or exploits vulnerabilities that let it take those computers over, it can then change all of the procedures that are supposed to monitor its behavior and limit its ability to take arbitrary actions on the internet without supervision.

If we lose those procedures [to cyberattack], then AIs working together could take any number of blatantly unwelcome, blatantly hostile takeover actions. Then it's moved beyond having to maintain secrecy and conspire. Things can escalate to physical weapons, takeover of social institutions, threats, things like that.

So, the critical thing to be watching is the software controls over the AI's motivations and activities. The point where things really go off the rails is when the hard power we once possessed over it is lost. Which can happen without us knowing it. Everything seems to be working well, we get happy reports. We think we're successfully aligning our AI, we think we're expanding its capabilities to do things like end disease. Countries concerned about the geopolitical military advantages are expanding their AI capabilities so they're not left behind. Meanwhile, all sorts of actions can be taken to set up a takeover of hard power over society.

The point where you lose the game is when you no longer have control over the AIs to stop them from taking all of these further incremental steps to takeover.

VI.   DWARKESH PATEL
San Francisco, the tech crowd, is paying attention to AI now. Who's going to be paying attention in 2026 and 2027?

LEOPOLD ASCHENBRENNER
Cofounder of Situational Awareness LP

National security is going to start paying a lot of attention to AI.•

DWARKESH PATEL

What is the immediate political reaction? Imagine if Xi Jinping sees the GPT-4 news and goes, "Look at the MMLU score on that. What are we doing about this, comrade?" What happens when he sees a remote worker replacement, and it has $100 billion in revenue?

LEOPOLD ASCHENBRENNER

The question is, when do the Chinese Communist Party and the American national security establishment realize that superintelligence is going to be absolutely decisive for national power? This is where the intelligence explosion stuff comes in. You have AGI. You have this drop-in remote worker that can replace you or me, at least for remote jobs. Fairly quickly, you turn the crank one or two more times, and you get a thing that's smarter than humans.

One of the first jobs to be automated is going to be an AI researcher or engineer. If you can automate AI research, things can start going very fast. Right now, there's already this trend of 0.5 orders of magnitude a year of algorithmic progress. At some point, you're going to have GPU fleets in the tens of millions or more for inference. You're going to be able to run 100 million human equivalents of these automated AI researchers. If you can do that, you can maybe make a decade's worth of ML research progress in a year. That's a 10x speed-up. You can make the jump to AI that is vastly smarter than humans within a couple of years.

That only broadens from there. After this initial acceleration of AI research, you do AI R&D on a bunch of other technologies. At this point, you have a billion superintelligent researchers, engineers, technicians, everything. They're superbly competent at all things. They're going to figure out robotics. We talked about that being a software problem. Well, you have a billion AI researchers smarter than the smartest human researchers in your cluster. They're going to be able to figure out robotics.

•   This episode was recorded in June 2024. In October, a White House memo stated, among other things, that "the United States Government must harness powerful AI, with appropriate safeguards, to achieve national security objectives." White House, "United States' Leadership in Artificial Intelligence."

If you play this picture forward, it's fairly unlike any other technology. A couple years of lead time could be utterly decisive in, say, military competition. If you look at the first Gulf War, Western coalition forces had a 100-to-1 kill-to-death ratio. They had better sensors on their tanks. They had better precision missiles, GPS, and stealth. They had a 20- to 30-year technological lead. They completely crushed the opposition.

Superintelligence applied to R&D, and the industrial explosion that comes from it—robots making a lot of materiel—could compress a century of technological progress into less than a decade. So a couple of years could mean a Gulf War-style advantage in military affairs, including a decisive advantage that preempts even nukes. How do you find nuclear stealth submarines? Right now, you have sensors and software to detect where they are. In the future, you might instead use millions or billions of mosquito-sized drones. They take out the nuclear submarines. They take out the mobile missile launchers. They take out the land- or air-based nukes.

It's potentially enormously destabilizing and enormously important for national power. At some point, people are going to realize that. When they do, it won't just be the AI researchers in charge. The CCP is going to have an all-out effort to infiltrate American AI labs. It'll involve billions of dollars, thousands of people, and the full force of the Ministry of State Security. The CCP is going to try to out-build us. They added as much power in the last decade as there is in the entire US electric grid. The 100 GW cluster, at least the power for it, is going to be a lot easier for them to get than us. It's going to be an extremely intense international competition.

VII.   DWARKESH PATEL

In your 2023 Senate testimony, you said that models are two to three years away from potentially enabling large-scale bioterrorism attacks.[102] Can you make that point more concrete without, obviously, giving the kind of information that would speed that up? What would that actually look like? Is it one-shotting• how to weaponize something, or do you have to fine-tune an open-source model?

•   "One-shotting," in this context, refers to creating a weapon on the first attempt using an ordinary LLM without any specialized post-training.

DARIO AMODEI
CEO of Anthropic

It'd be good to clarify this, because we did a blog post about this and people still didn't understand the point, or didn't understand what we'd done.[103]

Today, you can ask the models all kinds of things about biology and get them to say all kinds of scary things. But often those scary things are things that you could google. So I'm not particularly worried about that. I actually think it's an impediment to seeing the real danger.

We spent about six months working with world experts on how biological attacks happen, what you would need to conduct such an attack, and how we defend against such an attack. They worked very intensively on the entire workflow of trying to do a bad thing. It's not one shot. It's a long process. There are many steps to it. It's not like I asked the model for one page of information. Again, without going into any detail, there are some steps where you can just get the information on Google. Some steps are what I'd call missing—they're scattered across a bunch of textbooks, or they're not in any textbook. They're kind of implicit knowledge. I have to do this lab protocol, and what if I get it wrong? If this happens, then my temperature was too low. If that happened, I needed to add more of this particular reagent.

The models can't fill in the missing pieces yet, but we found that sometimes they can. Sometimes they still hallucinate, which is what's keeping us safe. But we saw enough signs of the models doing those key things well. If we look at state-of-the-art models and previous models and look at the trend line, there's every sign that two or three years from now, we're going to have a real problem.

DWARKESH PATEL

Yeah, especially on the log scale. You go from the model getting it right 1 in 100 times to 1 in 10 times, to...

DARIO AMODEI

GROK
A sudden improvement in model performance during training, reflecting the model's ability to generalize or learn an appropriate representation or algorithm for the task. Figuratively, a eureka moment. Originally coined by Robert Heinlein in *Stranger in a Strange Land*.

Exactly. I've seen many of these groks in my life. I was there when GPT-3 learned to do arithmetic, when GPT-2 learned to do regression a little bit above chance, when Claude got better on all these tests of helpfulness, honesty, harmlessness.[104] I've seen a lot of groks. This is not one I'm excited about, but I believe it's happening.

VIII.   JOE CARLSMITH
               Senior research analyst at Open Philanthropy

There are a bunch of ways this [relationship with powerful AI entities] can go. I'm not here to tell you 90 percent
doom or anything like that.

Here's the basic reason for concern. Imagine that we transition to a world in which we've created these beings that are
vastly more powerful than us. We've reached a point where
our continued empowerment is effectively dependent on their
motives. It is this vulnerability to what the AIs choose to do
[that worries me]. Do they choose to continue to empower
us, or do they choose to do something else?

                      DWARKESH PATEL
Or it's about the institutions that have been set up. I expect
the US government to protect me from AI, not because of
its motives but because of the system of incentives and
institutions and norms that has been set up.

                      JOE CARLSMITH
You can hope that will work. But there is a concern about
AI takeover scenarios along the spectrum of how much
power we *voluntarily* transfer to the AIs. How much of
our civilization did we hand to the AIs intentionally by
the time they took over, versus how much did they take
for themselves?

Some of the scariest scenarios are where it's a really fast
explosion. Maybe there wasn't even a lot of integration of
AI systems into the broader economy, but there's this really
intensive amount of superintelligence concentrated in a
single project. That's a scary scenario, partly because of
the speed and partly because of people not having time
to react.

There are intermediate scenarios where some things
were automated. Maybe people handed the military over
to the AIs, or we automated science, or we're doing all of
our cybersecurity with AIs. That's giving the AIs power
that they don't have to take from us.

Then there are worlds where we more fully transitioned
to a world run by AIs. In this scenario, humans voluntarily
did that. Maybe there were competitive pressures, but we
intentionally handed off huge portions of our civilization
to AIs. At that point, I think it's likely that humans have
a hard time understanding what's going on. A lot of stuff
is happening very fast. The police are automated. The
courts are automated.

I tend to think a little less about these scenarios because they're further down the line. When we look at technological adoption rates, adoption tends to be quite slow. But even in this case, it's intense. If humans have really lost their epistemic grip on the world, if they've handed the world to these systems... Even if we have laws and norms governing AI, I want us to have a really developed understanding of what's likely to happen in that circumstance before we go for it.

We may think there's going to be a bunch of intermediate time [for us to have these conversations]. But to me, it doesn't feel crazy that we might hand off the basic power to direct stuff to these automated systems [without fully understanding the implications]. Do we know enough about how that's going to go? Do we know enough about how the systems will be motivated to behave? I guess I'm trying to pump some general intuition [about what could go wrong], independent of these more specific scenarios.

Chapter 7
Explosion

What if intelligence makes
better intelligence?

The best answer to the question, "Will computers ever be as smart as humans?" is probably "Yes, but only briefly."
—Vernor Vinge, 2008

"Will AI take my job?" is a common refrain of the scaling era.• But this chapter addresses potential outcomes far beyond the automation of all human labor.

If AI systems can be used to improve AI systems, or AI research itself, it's easy to see how progress could accelerate enormously. Each successive AI would become increasingly adept at improving its successor, creating a compounding effect.

This feedback loop is already underway, albeit slowly. AI is reducing the costs of its own inputs, such as data center energy usage, training data, and new semiconductor designs, including the TPUs that Google uses to train its LLMs.•• It has taken over some research tasks, such as feature engineering, automatic differentiation, discovering key architectural details like activation functions and optimizers, and developing an improved algorithm for matrix multiplication.••• It's now pretty common to use an AI in evals to judge another AI's output.[105] The use of

FEATURE ENGINEERING
The practice of transforming raw input data to make it more informative in order to improve a model's predictions. This approach is largely obsolete, as neural networks can now learn better features autonomously.

AUTOMATIC DIFFERENTIATION
A computational method for performing calculus essential to many learning algorithms and simulations. While the backpropagation algorithm, which adjusts the weights of neural networks, already automated differentiation in neural networks, recent advances in automatic differentiation for more general functions and targets has been significant for science and engineering fields, where such calculus was previously done by hand.

• One memorable example is Senator Richard Blumenthal's question to Sam Altman during his Senate testimony: "I think you have said, in fact, and I'm gonna quote, 'Development of superhuman machine intelligence is probably the greatest threat to the continued existence of humanity.' You may have had in mind the effect on jobs, which is really my biggest nightmare in the long run." US Senate, "Oversight of AI."

•• In 2016, Google achieved a "15 percent reduction in overall PUE [power usage effectiveness] overhead after accounting for electrical losses and other non-cooling inefficiencies" by allowing a DeepMind system to redesign the cooling schedule for Google's data centers. It is important to note, however, that the applications mentioned here are not based on LLMs or Transformers and therefore do not qualify as literal, direct self-improvement. Evans and Gao, "AI Reduces Google Data Centre Cooling Bill"; Refuel, "Autolabel," https://github.com/refuel-ai/autolabel; Wang et al., "Want to Reduce Labeling Cost?"; Mirhoseini et al., "Graph Placement Methodology"; Goldie et al., "Addendum"; Roy et al., "PrefixRL."

••• Note that the AlphaTensor matrix multiplication algorithm works only on one very specific subtype of matrix. This is significant because matrix multiplication is the key operation in LLM training and inference. It involves taking two grids of numbers—for example, the weights of an LLM and the numerically encoded prompt provided by the user—and performing piecewise multiplication and summation to produce a third grid, the matrix product. Werbos, "Nonlinear Sensitivity Analysis"; Goyal et al., "Learning Activation Functions"; Chen et al., "Optimization Algorithms"; Fawzi et al., "Novel Algorithms."

LLM outputs for training LLMs, known as synthetic data, is still controversial, but it appears to complement human training data to some extent.• Training a successor system requires better chips, data, and algorithms—domains where existing AI has already made some progress.••

But these are indirect process improvements. When we talk about an intelligence explosion—AI accelerating AI research such that the rate of improvement explodes upward—we typically mean *direct* self-improvement: systems that can modify themselves or create their own successors.••• This, too, has some precedent, such as in the world-beating self-play RL systems discussed earlier in this book, like AlphaGo. So far, however, we've only seen this in closed-world problems, which have a fixed structure and simple objectives.••••

In this chapter, we discuss the evidence for and against a near-future intelligence explosion, our moral responsibility to detect and manage its impact, and the utopian and dystopian possibilities of achieving AGI.

I.   DWARKESH PATEL

Let's imagine we achieve transformative AI—the thing taking up a state's worth of power and many acres of GPUs. If the right model to get there is not "Delete everything, find new hyperparameters, start pretraining again" but rather "This exact same blob will continue getting smarter and smarter over time," that has all kinds of interesting implications for worries about takeover and alignment. If

•   When researchers were training Llama 3 to perform longer tasks, 0.1 percent of the total data used was synthetic; other models use much more than this. See also Feng et al., "Beyond Model Collapse"; Dubey et al., "Llama 3"; Sevilla et al., "Can AI Scaling Continue"; Lupidi et al., "Source2Synth."

••   In her interview, excerpted in this chapter, Ajeya Cotra notes that the nature of training could change dramatically. For instance, instead of training automatically with backpropagation, near-future AIs training their successors might directly intervene in the training process, for example by performing much more targeted credit assignment to weights.

•••   The term "intelligence explosion" was coined by the Bletchley cryptanalyst and pioneering statistician I.J. Good. It is also commonly referred to as the technological singularity. Good, "First Ultraintelligent Machine."

••••   See also OpenAI's o1, which appears to have been trained to prompt and reprompt itself, or algorithm distillation, an in-context method that automatically replaces RL algorithms with more sample-efficient versions. Laskin et al., "Reinforcement Learning with Algorithm Distillation."

SINGLETON
A system best modeled as a
unified decision-maker capable of forming and maintaining a world government.

there is this more coherent blob that can sustain itself, then that is a potential singleton AGI rather than the current situation, where we're constantly developing a bunch of different AGIs. What will it look like to deploy such a blob?

AJEYA COTRA
Senior program officer at Open Philanthropy

It really depends on the inference cost of running this blob. For the most powerful AI system out there, I can imagine its inference cost being roughly like the human brain or bigger. That might be $10 to $100 dollars an hour, which might not be worthwhile to shell out for many low-value tasks, although obviously it would be really valuable for frontier R&D.

In that world, you'd see a lot of distillation nonetheless, probably to distilled systems that still have an overkill amount of knowledge [for their allotted task]. If deep learning works as well as I expect it to work, it's probably not worth having a biology AI system that knows next to nothing about physics and English literature, because it's just not that expensive to have the general knowledge of physics and English literature in a Llama-sized package.• So I don't think you'll see the level of specialization we have for our physical tools, where we have one thing that only tills soil and a wrecking ball that only knocks down buildings. You'll see more generalization than that.

DWARKESH PATEL
If you can store every single piece of information in Wiki-pedia in less than 7 billion parameters, and if you have models like GPT-4 that are a trillion parameters, and if, in the future, you have 100 trillion-parameter models, what is going on with the rest of the parameters, the 100-trillion-minus-7-billion remainder that isn't storing facts?••

AJEYA COTRA
Most information or knowledge is not the kind of informa-tion or knowledge that you can lay out in Wikipedia. There's so much procedural knowledge. You have game-playing AI systems that get better and better as they get bigger and

• It would be about 405 billion parameters, or 820 GB on disk.

•• Seven billion 16-bit-precision parameters occupy approximately 13 GB on disk. For comparison, the compressed English-language text of Wikipedia is about 22 GB. Wikipedia, "Size of Wikipedia."

bigger, way outstripping the number of megabytes you need to store Wikipedia. Maybe in the future, videos or keystroke logging will be training fodder to develop models of the world that could be arbitrarily rich, heuristics about what to do in different situations one could encounter. These possible situations are exponentially greater than the amount of pieces of information on the web.

DWARKESH PATEL

Returning to the idea of the blob, one way to think about the blob is that we have this giant thing that's going to be bigger than the human brain and can be run with arbitrary amounts of inference. The blob has another advantage. Right now, Tesla, Google, or whoever have tens of thousands of middle managers because it's impossible for any one person to see every single input-output pair in Google. Imagine if the same model, with no principal-agent problem, is able to see everything that happens within Google. Would it be worth it for Google to spend $100 billion a year for this giga-model to consider how to make every single part of the company better? It probably would be! This is just a thing that no number of humans can do.

AJEYA COTRA

Yeah, I think the coordination benefits of AI systems are pretty underrated. Often, this is because futurism has tilted toward imagining one single entity bursting forth and causing harm. It's another thing you can tally onto the benefits of AI, along with not needing to sleep and being able to think much faster.

DWARKESH PATEL

Imagine that Elon Musk sees every role in Tesla. Whenever Tesla gets data from a mechanic in a Tesla dealership or the telemetry from a Tesla car, all of that goes straight to Elon's mind, and he's continuously learning from that. It does seem like if you could have the fluid intelligence of Elon plus this ability to coordinate that simply does not exist with humans... This is an argument for explosive growth that I hear less often. Then it seems very compelling.

Here's where I'm skeptical about this picture. I don't see how you go from current systems to the kind of thing that is able to do this fluid thinking.

AJEYA COTRA

There are two [possible] paths. One path is if we could somehow extract Elon Musk's thoughts from his brain. You

PRINCIPAL-AGENT PROBLEM
A common scenario where a customer (the principal) engages a professional (the agent) to act on their behalf. The problem arises when the agent's interests differ from those of the principal and the principal has limited information about the agent's work process or the quality of the outcome. Examples include medical treatment, appointing managers, or electing officials.

take a bunch of brilliant people who are really effective at their jobs, you figure out how they're thinking about what they're working on, and then you train the AI to predict the next *thought* step, not necessarily the next step in the world. I don't think we have concrete evidence of that path yet, except for maybe the early agents, where GPT-4 and Claude and so on can be hooked into these agent architectures and are able to ask themselves, "I wonder what this error message means." I'll call that the imitation learning path, or the supervised learning path. The other path is the RL path.

I think these paths can complement each other. On the RL path, we've seen superhuman capabilities in limited domains. On the imitation path, we've seen subhuman capabilities in broad domains. The AI industry is trying to get superhuman capabilities in broad, open-ended domains. The way we get there, in my mind, is to get as far as we can with imitation and then do as much RL as we need to.• If the combination of those two things is way too expensive, then AGI won't happen until we figure out a way to make it a lot cheaper and more sample efficient.

DWARKESH PATEL
I think we've hit on a crux, which is that both of these paths will be required to get there. Then the question is, will that happen in a few years, or will that happen in a century?

AJEYA COTRA
If you had to do billions or trillions of data points of really long-horizon RL, then that's more like the upper end of the Biological Anchors distribution•• and implies either quite long timelines—70 or 100 years—or some massively cost-saving new insight.

II.   DWARKESH PATEL
Give me the big-picture explanation of the feedback loops once you have something approaching human-level intelligence.

•   This references Yann LeCun's famous cake analogy: "If intelligence is a cake, the bulk of the cake is unsupervised learning, the icing on the cake is supervised learning, and the cherry on the cake is reinforcement learning." Articulated in 2016, the analogy anticipated the rough distribution of later LLM training compute. Synced, "Yann Lecun Cake Analogy."

••   The dataset in Cotra's "Draft Report on AI Timelines," published in September 2020.

CARL SHULMAN
Independent adviser to Open Philanthropy

AI research currently involves humans developing new computer chips, new software, and running larger training runs. It takes a lot of work to keep Moore's law chugging—when it was; it's slowing down now.[106] It takes a lot of work to develop things like Transformers, to develop a lot of the improvements to neural networks.

The core method that I want to highlight, which I think is underappreciated, is the idea of input-output curves. We can look at the increasing difficulty of improving chips. Each time you double the performance of computers, it's harder to do. As we approach physical limits, eventually it becomes impossible. But how much harder?

There's a paper called "Are Ideas Getting Harder to Find?" that was published a few years ago.[107] It covers a period where the productivity of computing went up a million-fold. You could get a million times the computing operations per second per dollar—a big change. But it got harder over time. The amount of investment and the labor force required to make those continuing advancements went up and up and up. It went up 18-fold over that period. Some take this to imply diminishing returns—things are just getting harder and harder, so that will be the end of progress eventually.

However, in a world where AI is doing the work, that doubling of computing performance translates pretty

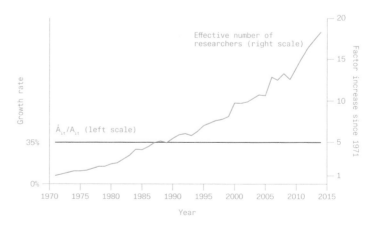

Figure 12. From 1970 to 2015, the growth rate of transistors per high-end chip remained a steady 35 percent per year. But at the end of the period, sustaining that exponential improvement required 18 times more researchers than in 1970. "The implication is that research productivity has fallen... an average rate of 6.8 percent per year," write Nicholas Bloom and coauthors. Bloom et al., "Are Ideas Getting Harder to Find?"

directly to a doubling, or better, of the effective labor supply. That is, if we use that million-fold compute increase to run intelligences that would replace human scientists and engineers, then the 18x increase in the human labor demands of the industry would be trivial. We're getting more than one doubling of the effective labor supply we need to cover the next doubled labor requirement. In the paper's dataset, it's over 4x—four to five doublings of compute per doubling of labor inputs. So when we double compute, we need somewhat more researchers, but a lot less than twice as many. We use up some of those doublings of compute on the increasing difficulty of further research, but most of them are left free to expedite the process.

So, if you double your labor force through AI, that's enough to get several doublings of compute. You use up one of them on meeting the increased demands from diminishing returns. The others can be used to accelerate the process. So your first doubling takes however many months, but your next doubling can take a fraction of that, the next doubling even less, and so on. At least insofar as the outputs you're generating—AI compute, in this story—are able to serve as all of the inputs. If you need other inputs, eventually those become a bottleneck, and you wind up more restricted.

DWARKESH PATEL
Can you explain the intuition that compute is a good proxy for the number of AI researchers?

CARL SHULMAN
So far, I've talked about hardware as an initial example because we had good data about a past period. You can also make improvements on the software side. When we think about an intelligence explosion, that can include AI doing work on making better hardware and software. But the basic idea for hardware is especially simple. If you have an AI worker that can substitute for a human and you have twice as many computers, you can run two separate instances of them. Then they can do two different jobs, manage two different machines, work on two different design problems. Now you can get more gains than you would by having two instances. We get improvements from using some of our compute not just to run more instances of the existing AI but also to train larger AIs.

There's hardware technology—how much you can get per dollar you spend on hardware—and there's software technology. The software can be copied freely. If you've got the software, it doesn't necessarily make that much sense to say, "We got you 100 Microsoft Windows." You can make as many copies as you need for whatever Microsoft will charge you. But for hardware, it's different. It matters how much we actually spend on the hardware at a given price. If we look at the changes that have been driving AI recently, that's the thing that is really off trend. We're spending tremendously more money on computer hardware for training big AI models.

DWARKESH PATEL

So, there's the investment in hardware, there's the hardware technology itself, and there's the software progress itself. AI is getting better because we're spending more money on it, because our hardware is getting better over time, and because we're developing better models or better adjustments to those models. Where is the loop here?

CARL SHULMAN

The work involved in designing new hardware and software is currently done by people. They use computer tools to assist them, but computer time is not the primary cost for Nvidia designing chips, nor for TSMC producing them, nor for ASML making the lithography equipment to serve TSMC. Even in AI software research, which has become quite compute intensive, we're still in the range where, at a place like DeepMind, salaries still cost more than the compute for the experiments. Although more recently, tremendously more expenditure went to compute relative to salaries.

ASML
A Dutch multinational that enjoys a near monopoly over the extreme ultraviolet lithography machines needed to produce leading chips. The most advanced chip manufacturers, including TSMC and Samsung, are dependent on access to ASML machines.

Take all the work being done by those humans. There are tens of thousands of people working at Nvidia designing AI GPUs. There are more than 70,000 people at TSMC, the leading producer of cutting-edge chips. There are a lot of additional people at companies like ASML who supply them with the tools they need. And then there's a company like DeepMind. From their public filings, they recently had 1,000 people. OpenAI has a few hundred people. Anthropic has less. If you add up things like Facebook AI Research, Google Brain, and other R&D, you get thousands or tens of thousands of people working on AI research [broadly construed].

We want to zoom in on those developing new methods rather than narrow applications. Inventing the Transformer definitely counts, but optimizing for some particular business's data cleaning? Probably not. The people doing that work on new methods are driving quite a lot of progress. What we observe in the growth of people relative to the growth of AI capabilities is that capabilities are doubling on a shorter time scale than the number of people required to do them.

On the software side, there's some work by [AI researcher] Tamay Besiroglu on this.[108] The doubling time of the workers driving the software advances is several years, whereas the doubling of effective compute from algorithmic progress is faster.

[AI research institute] Epoch AI's headline results for the rate of progress in hardware and software and growth in budgets are: For hardware, a doubling of hardware efficiency every two years.[109] It's possible AI hardware is a bit better than that, when you take into account specializations for AI workloads. For the growth of budgets, they find a doubling every six months, which is pretty tremendous relative to the historical rates.• On algorithmic progress, they find a doubling time of less than one year.•• When you combine all of these things, the growth of effective compute for training big AIs is pretty drastic.

III.   DWARKESH PATEL

Models of the intelligence explosion have so far come from economists. In their model, you replace the human AI researchers with a bunch of automated AI researchers that can speed up progress, make more AI researchers, and so make further progress. If I have a thousand agent Sholtos or agent Trentons, do I get an intelligence explosion? What does that look like to you?

SHOLTO DOUGLAS
Reinforcement learning infrastructure lead at Anthropic

One of the important bounding constraints here is compute. You *could* dramatically speed up AI research. It seems very clear to me that in the next couple of years, we'll have things that can do many of the software engineering tasks that I do on a day-to-day basis, and therefore can dramatically speed up my work, and therefore speed up the rate of progress.

But at the moment, the labs are somewhat compute bound. There are always more experiments you could run and more pieces of information you could gain. It's similar to how biology research is somewhat bound by experiment throughput. You need to run and culture the cells to get the

---

•   It's actually much faster for the largest frontier models. See Sevilla and Roldan, "Training Compute of Frontier AI Models."

••   Very roughly twice a year. Ho et al., "Algorithmic Progress in Language Models."

information. So compute will be at least a short-term planning constraint. Obviously, Sam Altman is trying to raise $7 trillion to buy chips, and there's going to be a lot more compute in the future as everyone is heavily ramping.•

TRENTON BRICKEN
Interpretability researcher at Anthropic

We need a few more nines of reliability for it to be really useful and trustworthy. We need context lengths that are super long and very cheap. If I'm working in our codebase, I can really only get Claude to write small modules for me right now. But it's very plausible that within the next few years, or even sooner, it will automate most of my tasks.

DWARKESH PATEL

By the time the intelligence explosion is happening, we will have models that are two to four orders of magnitude bigger in terms of effective compute. In this version of the intelligence explosion, with this idea that you can run experiments faster, you're having to retrain that model. The recursive self-improvement is different from what might have been imagined 20 years ago, where you could just rewrite the code. You have to train a new model, and that's really expensive. As you keep making these models orders of magnitude bigger, doesn't that training requirement dampen the possibility of a recursive self-improvement type of intelligence explosion?

SHOLTO DOUGLAS

It's definitely going to act as a braking mechanism. I agree that what we're making today looks very different from what people imagined it would look like 20 years ago. It's not going to be able to write the same code to be really smart, because it needs to train itself. The code itself is typically quite simple, typically really small and self-contained.

John Carmack [founder of AGI company Keen Technologies] had this nice phrase: It's the first time in history where you can plausibly imagine writing AI with 10,000 lines of code.[110] That does seem plausible when you pare training codebases down to the limit.

IV.   DWARKESH PATEL
How likely does a forcible AI takeover seem?

•   This statistic is often misunderstood. Altman was actually
        referring to $7 trillion as the total global investment required.

CARL SHULMAN

Before the deep learning revolution, I might have said 10 percent. I expected us to have a lot more time to build movements and prepare for these [alignment and governance] problems in advance. But that was only some 15 years ago. We didn't get 40 or 50 years as I'd hoped. The situation is moving very rapidly now. At this point, depending on the day, I might say 25 or 20 percent.

DWARKESH PATEL

Given the very concrete ways in which you explain how a takeover could happen, I'm surprised you're not more pessimistic. Why not?

CARL SHULMAN

A lot of it is driven by intelligence explosion dynamics. Our attempts to do alignment have to take place in a very short time window, because if a [capability required to automate AI] safety emerges only when an AI has near-human-level intelligence, that's potentially deep into an intelligence explosion. You have to do things very quickly. Handling that transition may be the scariest period of human history, although it also has the potential to be amazing.

The reason why I think we have a relatively good chance of handling the transition is twofold. One is that as we approach that kind of AI capability, we're approaching it with weaker systems, like current predictive models, which start off with less situational awareness. [In human evolution] humans can develop a number of different motivational structures in response to simple reward signals, but we often wind up with things pointed in roughly the right direction. With respect to food, for example, our hunger drive is pretty effective, although it has weaknesses. We get to apply much more selection pressure [to AI] than was the case for ancient humans by actively generating situations where the [desirable and undesirable motivations] might come apart—situations where a bit of a dishonest tendency or a bit of motivation to attempt a takeover or subvert the reward process gets exposed.•

A perfect AI that can always figure out exactly when it would get caught and when it wouldn't might navigate that with a motivation of only conditional honesty or only conditional loyalties. But for systems that are limited in their ability to reliably determine when they can get away with things and when they can't, including our efforts to actively construct those [red-team test] situations, and including our efforts to use interpretability methods to create neural lie detectors, it's quite a challenging situation to develop those motives.

We don't know when in the process those motives might develop. If the really bad sorts of motivations develop relatively later in the training process, at least with all of our countermeasures, then by that time we may have plenty of ability to extract AI assistance in further strengthening the quality of our adversarial examples, the strength of our neural lie detectors, the experiments we can use to reveal, elicit, and distinguish between different kinds of reward hacking tendencies and motivations.

We may have systems that have not developed bad motivations in the first place, and we may be able to use them a lot in developing the incrementally better systems in a safe way. We may be able to develop methods of interpretability, seeing how different training methods work to create them, even if some of the early systems do develop these bad motivations. If we're able to detect that, experiment, and find a way to get away from that, then we can win even if these hostile motivations develop early.

DWARKESH PATEL
Is there any hope that having leverage over the complex global supply chains that advanced rogue AIs would initially rely on to accomplish their goals would make it easy to disrupt their behavior?

CARL SHULMAN
That doesn't help much in the central case where the AIs are secretly misaligned and global supply chains are constructing everything they need for fully automated infrastructure. In cases where AIs tip their hands at an earlier point, then yes, it seems like it adds some constraints. In particular, these large server farms are identifiable and more vulnerable.

It seems to me, though, that the main protective effect of that centralized supply chain is that it provides an

REWARD HACKING (or SPECIFICATION GAMING) When an AI system exploits loopholes or bugs to maximize its reward function in ways that diverge from the intended task. This phenomenon is common even in simple RL agents.

• Here, Shulman is referring to adversarial testing or red-teaming: testing strategies designed to identify scenarios where an AI fails at a task (for example, by hiding its true motivations or being reinforced toward dishonesty) with the goal of training the model to avoid such failures in the future. As an analogy, Doritos are sometimes described as a "superstimulus" for humans because they exploit our evolved sense of correlation between taste, nutrition, and caloric value by being delicious but not nutritious. The idea is that if humans could be trained like AIs, we could identify edge cases like Doritos and apply negative reinforcement to "patch" them.

opportunity for global regulation to restrict the unsafe racing forward without adequate understanding of AI systems before this nightmarish process could get in motion.

DWARKESH PATEL

How about the idea that if this AI was trained with a $100 billion training run, it's going to have many trillions of parameters, and it would be hard for that to be stored on one gaming GPU somewhere?

CARL SHULMAN

Storage is cheap. Hard disks are cheap.

DWARKESH PATEL

But it would need a GPU to run inference.

CARL SHULMAN

While humans have similar quantities of memory and operations per second, GPUs have very high numbers of FLOPs compared to the high-bandwidth memory on the chips. The ratio can be, like, 1,000-to-1. The leading Nvidia chips may do hundreds of teraflops but only have 160 GB of high-bandwidth memory. That is a limitation. If you're trying to fit a model whose weights take 80 TB, you'd have to have a large number of those chips for the model to work on many tasks at once and have data parallelism. It would be a restriction for a model that big to be stored on one GPU.

DATA PARALLELISM
The practice of splitting data across multiple machines and training on multiple batches of data simultaneously. This approach is challenging in practice because training updates need to be synchronized. Here, Shulman is instead talking about inference parallelism: having one model work on many subtasks at once.

Now, there are things that could be done to change that with the incredible level of software advancement from the intelligence explosion. Engineers can surely distill a lot of capabilities into smaller models by rearchitecting things. They can make new chips with different properties. But yes, the most vulnerable phases are going to be the earliest. These chips are relatively identifiable early on, relatively vulnerable, which would be a reason why you might expect this kind of AI takeover to initially involve secrecy, if that was possible.

V.   DWARKESH PATEL

One concern people have is, if we're going to have very powerful systems capable of an intelligence explosion, it would be bad to live in a world where one group, whether it's a company or a government, is in charge of it. What might be best is to have pre-intelligence-explosion systems as broadly deployed as possible so there's a balance of power. We want to make sure one company doesn't control the universe. So should we be open-sourcing the Claude 6–level models?

JARED KAPLAN

Cofounder of Anthropic

I completely agree that one company shouldn't get to decide to let it rip and trigger the intelligence explosion, or even systems less powerful than that, at broadly human level. That is a national and international conversation.

My view is that we don't have a lot of time, but the goal would be to agree that systems beyond a certain capability level—systems that could make terrorists much more effective at building bioweapons, for example—should have constraints on how they're deployed, how they're used, who they're available to. But those constraints should apply to everyone. They should apply to Anthropic. They should apply to other AI developers. They should apply to open-source models.

Open source is great because it makes it much easier for all sorts of different researchers to study interpretability and alignment and understand how AI works. But there should be some regulation, agreement, treaty, whatever you want to call it, that sets limits on capabilities for systems that could be misused, that might be dangerous, and then require certain mitigations for those risks.

VI.   DWARKESH PATEL

Here's a claim somebody could make. If these things hang around human level, and if they're trained the way they currently are, recursive self-improvement is much less likely because they're at human-level intelligence. It's not a matter of optimizing some for loops, or something, because they've got to train another billion-dollar run to scale up. In that scenario, that kind of recursive self-intelligence is less likely. How do you respond?

ELIEZER YUDKOWSKY

Cofounder of the Machine Intelligence Research Institute

At some point, they get smart enough that they can roll their own AI systems and are better at it than humans. That is the point at which you definitely start to see foom. Foom could start before then for some reason, but we are not yet at the point where you would obviously see foom.

FOOM
Onomatopoeia for an intelligence explosion.

DWARKESH PATEL

Why doesn't the fact that they're going to be around human level for a while increase your odds of human survival?

Because that gives us more time to align them. Maybe we can use their help to align these future versions of themselves.

ELIEZER YUDKOWSKY

Having AI do your AI alignment homework for you is the nightmare application for alignment. Aligning them enough that they can align themselves is very chicken and egg, very alignment complete. If you have these things trying to solve alignment for you, they need to understand AI design. If they're a large language model, they're very good at human psychology, because predicting the next thing you'll do is their entire deal. Also game theory and computer security and adversarial situations, and thinking in detail about AI failure scenarios in order to prevent them. There are just so many dangerous domains [an AI would] have to operate in to do alignment.

ALIGNMENT COMPLETE
Referring to something that is as difficult to achieve as alignment itself.

VII. DWARKESH PATEL

Let's talk about what a post-AGI future looks like. I'm guessing you're working 80-hour weeks toward this grand goal that you're really obsessed with. Are you going to be satisfied in a world where you're basically living in an AI retirement home? What are you personally doing after AGI comes?

ILYA SUTSKEVER
Cofounder of Safe Superintelligence Inc.

The question of what I'll be doing or others will be doing after AGI is very tricky. Where will people find meaning? But that's a question AI could help us with. I imagine we will be able to become more enlightened because we interact with an AGI. It will help us see the world more correctly and become better on the inside as a result of interacting with it. Imagine talking to the best meditation teacher in history. That will be a helpful thing.

PROCESS NODE
A semiconductor manufacturing process, usually expressed in nanometers or angstroms. Originally, the term referred to the length of a single gate on a single transistor, but it has since become a marketing designation with no clear referent. The smaller the transistor, the faster and more efficient it is. Currently, the most advanced companies can produce so-called 3nm nodes.

VIII. DWARKESH PATEL

Would it be possible for the intelligence explosion to happen without any hardware progress? If hardware progress stopped, would this feedback loop still be able to produce some explosion with software alone?

CARL SHULMAN

If the technology is frozen—no more nodes, Moore's law is over—at that point, gains come from constructing more chips. You could still realize economies of scale there.

Right now, a chipmaker has to amortize the R&D cost

of developing the chip. You build a fab, and its peak profits come in the few years that the chips are cutting edge. Later on, as the cost of compute exponentially falls, you keep the fab open because you can still make some money, given that it's already built. But the profits the fab will make are relatively front-loaded. So, in a world where Moore's law ends, you wind up with very long production runs, where you can keep making chips that stay at the cutting edge and your R&D costs get amortized over a much larger base. The R&D basically drops out of the price, and then you get some economies of scale from making so many fabs.

This is applicable in general across industries. When you produce a lot more, the costs fall. ASML has many incredibly exotic suppliers that make one bizarre part out of the thousands of parts in one of their machines that you can't get anywhere else. They don't have standardized equipment because this is the only use for it. In a world where we're making 10 or 100 times as many chips at the current node, they would benefit from scale economies. All of that would become more industrialized.

Combine all of those things, and you get a decline in the capital costs of buying a chip, but not in the energy costs of running the chip. Right now, energy costs are a minority of the total cost, but they're not trivial. It passed 1 percent a while ago, and they're inching up toward 10 percent and beyond. You can maybe get another order-of-magnitude cost decrease from getting really efficient at plant construction, but energy would still be a limiting factor once actual chip improvement ends.

IX.   DWARKESH PATEL
What do you think the odds of value lock-in are given transformative AI?

HOLDEN KARNOFSKY
Member of technical staff at Anthropic

Lock-in is the possibility that we could end up with a very stable civilization. Throughout history, when someone takes charge of a government and they're very powerful and very bad, it's generally considered to be temporary. The world is dynamic and tends not to stay any particular way globally. If someone is running a country in a really cruel, corrupt way, at some point they're going to get old and die, and someone else is going to take over. That person will probably be different from them. Furthermore, the world is changing all the time. There are new technologies, new things are possible, there are new ideas. The most powerful country today might not be the most powerful tomorrow. The people in power today might not be the ones in power tomorrow. This gets us used to the idea that everything is temporary and everything changes.

A point I make in the Most Important Century series[111] is that you can imagine a level of technological development where there just aren't new things to find. There isn't a lot of new growth. People aren't dying, because it's medically possible for people not to age or die. You can imagine a lot of the sources of dynamism in the world going away if we had enough technology. You could imagine a government that was able to surveil everyone, which is not something you can do now, with a dictator who doesn't age or die, who knows everything going on, who's able to respond to everything. You could imagine that world being completely stable.

This is a very scary thought. It's something we have to be mindful of. If the rate of technological progress speeds up a lot, we could quickly get to a world that doesn't have a lot more dynamism.

What are the odds of this? It's very hard to put a probability on it. But if you imagine that we're going to get this explosion in scientific and technological advancement, you have to take pretty seriously the idea that we could end up hitting a wall. We could have these very stable societies.

What does that seriously mean in terms of probability? I don't know. A quarter, a third, a half, something like that. I'm making up numbers. It's serious enough to think about as something that affects the stakes of what we're talking about.

X.   DWARKESH PATEL

We're coming at this conversation from the perspective that there's going to be a race with China, and the US must win. But I want to step back to a galactic perspective. Humanity is developing AGI. To our superintelligent, Jupiter-brain descendants, China will be a distant memory. America too. Shouldn't the initial approach be more cooperative? Why immediately rush into it from a hawkish, competitive perspective?

LEOPOLD ASCHENBRENNER
Cofounder of Situational Awareness LP

A lot of the stuff in the Situational Awareness series is primarily descriptive. In some ideal world, it's all merry-go-round and cooperation. The issue in particular is, *can we make a deal? Can we make an international treaty?*

BREAKOUT
A term of art in security
studies referring to the time
between the initiation of
an attack (or defection) and
its success. As an example,
it can describe the time
required for a hostile actor
to produce enough highly
enriched uranium for a single
nuclear weapon.

This relates to the stability of international arms control agreements. We did very successful arms control on nuclear weapons in the 1980s. The reason it was successful is because the new equilibrium was stable. You go down from 60,000 nukes to 10,000 nukes, or whatever. When you already have 10,000 nukes, breakout basically doesn't matter that much. Suppose the other guy now tried to make 20,000 nukes. Who cares? It's still mutually assured destruction. Suppose a rogue state went from zero nukes to one nuke. Who cares? We still have way more nukes than you. It's still not ideal for destabilization.

It'd be very different if the arms control target had been zero nukes. At zero nukes, you just need one rogue state to make one nuke and the whole thing is destabilized. Breakout is very easy. Your adversary state starts making nukes. When you're going to very low levels of arms, or when you're in a very dynamic technological situation, arms control is really tough because breakout is easy.

There are some other stories about this in the 1920s and 1930s. All of the European states had disarmed. Germany did this crash program to build the Luftwaffe. They were able to massively destabilize Europe because they were the first [to do that]. They were able to pretty easily build a [dominant] modern air force because the other nations didn't really have one.

The issue with AGI and superintelligence is the explosiveness of it. If you have an intelligence explosion, you're able to go from AGI to superintelligence. That superintelligence is decisive, because you'll develop some crazy WMD or you'll have some super hacking ability that lets you completely deactivate the enemy arsenal.

Suppose you're trying to put in a break. We're both going to cooperate, we're going to go slower on the cusp of AGI. In that situation, there is going to be such an enormous incentive to race ahead, to break out and do the intelligence explosion. If we can get three months ahead, we win. That makes any arms control agreement very unstable in a close situation.

DWARKESH PATEL

This is analogous to a debate I had with Richard Rhodes [author of *The Making of the Atomic Bomb*] where he argued for nuclear disarmament. He said that if some country tried to break out and started developing nuclear weapons, the six months it would take to do that is enough time to get international consensus and invade the country and prevent them from getting nukes. I thought that was not a stable equilibrium.

In this case, maybe it's a bit easier because you have AGI, so you can monitor the other person's cluster, or something. You can see the data centers from space. You can see the energy draw they're getting. Also, unlike nukes, the data centers are fixed in place. Obviously, you can have nukes in submarines, planes, bunkers, mountains. We can blow up a 100 GW data center if we're concerned. That's very vulnerable.

LEOPOLD ASCHENBRENNER

That gets to the insane vulnerability and volatility of this period post-superintelligence. You have the intelligence explosion. You have these vastly superhuman things running on your cluster. You haven't done the industrial explosion yet. You don't have robots yet. You haven't covered the desert in robot factories yet. That is the crazy moment.

Say the United States is ahead and the CCP is somewhat behind. There's actually an enormous incentive for the CCP to launch a first strike, if they can take out your data center. They know you're about to have this command, a decisive lead. They know that if they can just take out this data center, they can stop it. They might get desperate.

We're going to get into a position that's going to be pretty hard to defend early on. We're basically going to be in a position where we're protecting data centers with the threat of nuclear retaliation. Maybe that sounds kind of crazy.

XI.   DWARKESH PATEL

What do the next few years until transformative AI look like? Do you expect that it will be a government project by default? Who controls it? What level of secrecy will it have?

AJEYA COTRA

In the next six months [by early 2025], AI models will become more reliable personal assistants or research assistants, useful at one- to two-hour-long tasks. I expect that to drive another round of people getting excited about AI and another order-of-magnitude increase in power users relying on these systems. That will naturally draw attention to the field. I am pretty uncertain about whether AI will ever get securitized and nationalized in the way that Leopold predicts in his Situational Awareness manifesto. Right now, the frame is that it's a commercial technology.

The default path I imagine is increasing government scrutiny over time. Maybe the government will become an increasingly big customer of AI companies, in the way the defense establishment is a big customer of Boeing. But there will probably still be a separation [between AI companies and the government], and the company will sell to a bunch of other people besides the government. The company is regulated by the government, and also maybe soft-regulated by the relationships they have with people in government who don't necessarily have hard legal power over them. I imagine that's the path of least resistance.

DWARKESH PATEL

Is it a good thing or a bad thing?

AJEYA COTRA

I personally find it a little more comforting than the fully securitized path. The frame everyone has in mind is that AI is a weapon, so everything gets very locked down, very secret, with a tighter connection between the government and the companies. But a lot of people in the safety world disagree with that.

DWARKESH PATEL

The key part of this picture is that we don't know what should be done about AGI. It's very easy to be wrong about the best thing to do. We have the further disadvantage that it's not clear to all of the people trying to make decisions what is actually going on. It's a cursed epistemic situation. So how should it be done?

AJEYA COTRA

In my dream world, the science is such that it's not that easy to radically zoom through levels of intelligence. Like, even if you use AI to automate AI R&D, there's some time before this AI-automating-AI loop really gets going. If that's the case, then there are a lot of opportunities for society to both formally and culturally regulate different streams in different ways and at different rates. That's the vision that some of the open-source community wants. Don't regulate the source product, regulate the different applications.

My fear is that that's not how it is. Instead, I worry that a lot of really powerful things will come really quickly, and there will be this winner-take-all dynamic where a firm that's only six months ahead can get an irreversible lead and a bunch of power over everyone else. If we're in that world, our best hope is for people to see that ahead of time, before we go off the precipice.

DWARKESH PATEL

We want to win against China because we don't want a top-down authoritarian system to win the AI race. But if the way to beat them is for the most important technology to be controlled by a top-down government, aren't we becoming the same thing?

LEOPOLD ASCHENBRENNER

There's a misconception that AGI development will be a beautiful, decentralized thing, a giddy community of coders collaborating. That's not how it's going to look. It's going to be trained on a $100 billion or trillion-dollar cluster. Not many people will have one of those. Moore's law is really slow. AI chips are getting better, but the $100 billion computer won't cost $1,000 within your lifetime. So there are going to be two or three big players in the private world.

You talk about the enormous power that superintelligence and the government will have. It's pretty plausible that in the alternative world, one AI company will have that power. Say OpenAI has a six-month lead. You're talking about the most powerful weapon ever. So you're making a radical bet on a private company CEO as the benevolent dictator.

DWARKESH PATEL

Not necessarily. Like any other thing that's privatized, we don't count on CEOs being benevolent. Think of someone who manufactures industrial fertilizer. They could probably blow up Washington, DC. Right now, a lot of private actors control vital resources like the water supply. We count on cooperation and market-based incentives to maintain a balance of power. We have a lot of historical evidence that this works best.

LEOPOLD ASCHENBRENNER

I'm not sure this analogy holds. The key differences are speed and offense–defense balance issues. It's like compressing the 20th century into a few years. That is incredibly scary because of the rapid advancement in destructive technology and military advancements. That's where a government project is necessary. If we can make it through that, the situation stabilizes.

Bioweapons are a huge issue initially. An attacker can create 1,000 different synthetic viruses and spread them. It's hard to defend against each. Maybe at some point, you figure out a universal defense against every possible virus, and then you're in a stable situation again on the offense–defense balance. Or think about planes. You restrict certain capabilities that the private sector isn't allowed to have, and then you can let the civilian uses run free.

One company having all the power would be unprecedented. The industrial fertilizer manufacturer cannot overthrow the US government, but it's quite plausible that the AI company with superintelligence can. If there are two or three companies [building superintelligent AI], then it's a crazy race between these companies. Demis Hassabis and Sam Altman would be like, "I don't want to let the other one win." They're both developing their nuclear arsenals and robots. Come on, the government is not going to let these people do that.

The other issue is that if it's two or three [companies building superintelligence], it won't just be two or three [American companies]. It'll be China, Russia, and North Korea too. No way will the private lab world have good enough security [to defend against nation-state actors].

ARTIFICIAL SUPERINTELLIGENCE (ASI) Another term for AGI, referring specifically to AI systems that surpass human-level capabilities.

DWARKESH PATEL

If you can cause a coup with ASI, the same capabilities are going to be true of the government project too. If the takeoff is slower than anticipated, I prefer the private companies. In no part of this matrix is it obviously true that the government-led project is better.

LEOPOLD ASCHENBRENNER

In some sense, my argument is a Burkean one. American checks and balances have held for over 200 years, through crazy technological revolutions.

DWARKESH PATEL

The private–public balance of power has also held for hundreds of years.

LEOPOLD ASCHENBRENNER

But why has it held? Because the government had the biggest guns. Never has a single CEO or a random nonprofit board had the ability to launch nukes. What is the track record of government checks and balances versus the track record of private company checks and balances? Well, the AI labs' first stress test went really badly.•

Even worse, in the [non-state-operated ASI] world, it's

• Referring to the November 2023 coup and counter-coup at OpenAI, in which the nonprofit board fired CEO Sam Altman but reinstated him after intense blowback. Five of the six members of the board subsequently resigned.

private companies versus the CCP, so the CCP will instantly have all of the tech [via espionage]. The companies probably won't have good enough internal controls. It's not just the CEO [you need to worry about] but rogue employees who can use these superintelligences to do whatever they want.

DWARKESH PATEL
And rogue employees won't exist in the government project?

LEOPOLD ASCHENBRENNER
The government has decades of experience and actually cares about this stuff. They deal with nukes and really powerful technology.

Also, what are the checks and balances in the government world? First, it's important to have an international coalition. The inner tier is modeled on the Quebec Agreement: Churchill and Roosevelt agreeing to pool efforts on nukes but not using them against each other or anyone else without consent. In AI, you could bring in the UK with DeepMind, Southeast Asian states with the chip supply chain, and more NATO allies with talent and industrial resources. So you would have checks and balances from having more countries at the table.

Separately, you have the second tier of coalitions, an Atoms for Peace thing.• You go to countries including the UAE and make a deal similar to the Non-Proliferation Treaty. They're not allowed to do crazy military stuff, but we'll share civilian applications. We'll help them and share the benefits, creating a new post-superintelligence world order.

Then there are the US checks and balances. You have Congress, different factions of the government, and the courts. Congress will have to be involved to appropriate trillions of dollars. Ideally, Congress will need to confirm whoever's running this. I expect the First Amendment to remain really important.

These institutions have withstood the test of time in a powerful way. This is why alignment is important. You program AIs to follow the US Constitution. The military works because generals are not allowed to follow unlawful or unconstitutional orders. We can have the same thing for the AIs.

• A US Cold War program that supported the building of civilian reactors in various countries as an alternative to state-owned military nuclear projects. The first nuclear reactors in Israel and Pakistan were supplied by Atoms for Peace; both countries are now nuclear armed. Cohen and Burr, "Israel's Secret Nuclear Project"; Khan and Khan, "Pakistani Nuclear Program."

DWARKESH PATEL

The argument that these institutions have held up well
historically is flawed. They've actually almost broken a
bunch of times.

LEOPOLD ASCHENBRENNER

But they held up. They didn't break the first time they
were tested. There are institutions, constitutions, legal
restraints, courts, and checks and balances. The crazy
bet is betting on a private company CEO.

DWARKESH PATEL

What's wrong with a scenario where multiple companies
are going for ASI? The AI is still broadly deployed. Alignment works. The system-level prompt is that it can't help
people make bioweapons, or something.

LEOPOLD ASCHENBRENNER

I expect AIs to be broadly deployed even if it's a government project. The Metas of the world open-sourcing AIs
that are two years behind is super valuable. There will be
some question about whether the offense-defense balance is fine, whether open-sourcing two-year-old AIs is
fine. Maybe there are restrictions on the most extreme
dual-use capabilities, like not letting private companies
sell weapons. That's great and will help with diffusion.
After the government project, there will be an initial tense
period. Hopefully it stabilizes. Then, like Boeing, [the
private companies] will go out and do all of the flourish-
ing civilian applications, like nuclear energy. The civilian
applications will have their day.

DWARKESH PATEL

So then Google, because they got the contract from the
government, will control the ASI?

LEOPOLD ASCHENBRENNER

It will be the same companies that would be doing it anyway.
In this world, they're just contracting with the government
or are DPA'd so that their compute goes to the government.•

• The 1950 Defense Production Act (DPA) grants the US president
  sweeping powers over any part of the economy deemed critical
  to national defense. These include enforcing mandatory contracts, creating nationalized supply agencies, and implementing
  price and supply controls. FEMA, "Defense Production Act."

DWARKESH PATEL

After you get the ASI and we're building the robot armies...

LEOPOLD ASCHENBRENNER

Only the government will get to build robot armies.

DWARKESH PATEL

Now I'm worried.

LEOPOLD ASCHENBRENNER

That's what we do with nukes today. The government has the biggest guns. The way we regulate is through institutions, constitutions, legal restraints, courts.

Either you barely scrape ahead but you're in a fever struggle, proliferating WMDs. That's enormously dangerous for alignment, because you're in a race and you can't take six months to get alignment right. The alternative is not bundling efforts to win the race against authoritarian powers.

I don't *like* it. I wish we could use the ASI to cure diseases and do all the good in the world. But it's my prediction that in the endgame, what is at stake is not just cool products but whether liberal democracy survives or the CCP survives.

In terms of whether researchers will sign up for it, they wouldn't sign up for it today. It would seem crazy to people. I don't think anyone has really gotten in front of these people and said, "Look, what you're building is the most important thing for the national security of the United States, for the future of the free world, and for whether we have another century ahead of it." It's not just about DeepMind, or whatever, it's about these really important things.

DWARKESH PATEL

When you say that we need to pursue this project for the free world to survive, it sounds similar to World War II. World War II is a sad story not only because it happened, but because Britain originally went in to protect Poland. At the end, the USSR ends up occupying half of Europe, including Poland. We might look back on this with the same twisted irony as Britain going into World War II to protect Poland.

LEOPOLD ASCHENBRENNER

There will be a lot of unfortunate things that happen. I'm just hoping we make it through.

Safety, including alignment and the control of new WMDs, is [objectively] important. I'm not convinced there's another path. Alignment during the intelligence explosion is not a yearslong bureaucratic process. It's more like a fog of war: "Is it safe to do the next OOM? We're three OOMs into the intelligence explosion and we don't fully understand what's happening. Our generalization scaling curves don't look great. Some automated AI researchers say it's fine, but we don't quite trust them. Should we go ahead?" Meanwhile, China might steal the weights or deploy a robot army.

It's a crazy situation, relying more on a sane chain of command than a deliberative regulatory scheme. Private companies claim they'll do safety, but it's rough in a commercial race, especially for startups. Startups are startups. They aren't fit to handle WMDs.

XIII.   DWARKESH PATEL

Misuse and misalignment are both big problems. But in the long run, say, 30 years down the line, which do you think will be considered a bigger problem?

DARIO AMODEI
CEO of Anthropic

I think it's going to be much less than 30 years, and I'm worried about both. If you have a model that could, in theory, take over the world on its own, and if you were able to control that model, then it follows pretty simply that if a model was following the wishes of some small subset of people and not others, those people could use it to take over the world on their behalf.

DWARKESH PATEL

Some would say that you're already assuming an optimistic scenario here, because in this scenario you've at least figured out how to align the model with *someone*, so now you just need to make sure it's aligned with the good guys instead. Why do you think that you could get to the point where it's even aligned with the bad guys? We haven't solved this yet.

DARIO AMODEI

Any plan that actually succeeds, regardless of how hard misalignment is to solve, is going to need to solve misuse as well as misalignment. As the AI models get better faster and faster, they're going to create big problems around the balance of power between countries, and whether it's possible for a single individual to do something bad that's hard for everyone else to stop.

DWARKESH PATEL

If misuse doesn't happen and the "right" people have the superhuman model, what does that look like? Who are the right people? Who is controlling the model five years from now?

DARIO AMODEI

These things are powerful enough that the field is going to require substantial involvement from some kind of government or assembly of government bodies. It's too powerful [for one company or government]. There needs to be some kind of legitimate process for managing this technology, which includes the people building it, democratically elected authorities, and all of the individuals who will be affected by it.

DWARKESH PATEL

But what does that look like? If it's not the case that you just hand it to whoever the president is at the time, what does the body look like?

DARIO AMODEI

The honest fact is that we're figuring this out as we go along. We should try things and experiment with them with less powerful versions of the technology. We need to figure this out in time. But it's not really the kind of thing you can know in advance.

Chapter 8
Timelines

When will we get AGI? •

When our most advanced tech was fire, people thought life was like fire, and they were right. Our energy comes from an oxidation reaction.

When our most advanced tech was steam engines, people thought life was like a steam engine, and they were right. Our bodies have a pump at the center that drives them...

When our most advanced tech was computers, people thought life was like a computer, and they were right. Our brains send patterns of electrical impulses through a network...

When our most advanced tech was LLMs, people thought we were like LLMs, and they were right. Fill in the blank.
—Ryan Moulton, 2023[112]

I.   SHANE LEGG
Cofounder and chief AGI scientist at Google DeepMind

I think there's a 50 percent chance we have AGI by 2028. Now, that's just a 50 percent chance. I'm sure what's going to happen is 2029 will pass and someone is going to say, "Shane, you were wrong!" Come on, I said a 50 percent chance.

II.   DEMIS HASSABIS
Cofounder and CEO of Google DeepMind

When we started DeepMind back in 2010, we thought of it as a 20-year project. I think we're on track [for AGI in 2030]. Which is kind of amazing for a 20-year project, because usually they're *always* 20 years away.

III.   ILYA SUTSKEVER
Cofounder of Safe Superintelligence Inc.

How long until AGI? It's a hard question to answer. I hesitate to give you a number.

IV.   DARIO AMODEI
CEO of Anthropic

In terms of someone looking at the model and, even if they talk to it for an hour or so, determining that it's basically

* It is important to note that each interviewee uses a different definition of AGI.

like a generally well-educated human, that could be not very far away at all. That could happen in two or three years [2025 or 2026].

V.   JARED KAPLAN
                     Cofounder of Anthropic

At this point, I hold out 10 to 30 percent for the scenario that I'm just kind of nuts, and I'll understand I was making the wrong assumptions in five or 10 years. But it definitely does feel like we'll have human-level AI given 10^{29} or 10^{30} training FLOPs, so maybe late this decade [2029]. That's an extremely daunting thought.

VI.   LEOPOLD ASCHENBRENNER
               Cofounder of Situational Awareness LP

By 2027 or 2028, it'll get as smart as the smartest experts. It'll almost be like a drop-in remote worker. Also, it's unlikely, but there are worlds where we get AGI next year [in 2025].

VII.   HOLDEN KARNOFSKY
                Member of technical staff at Anthropic

You could imagine the world changing incredibly quickly and incredibly dramatically. It looks reasonably likely—in my opinion, more than 50–50—that this century will see AI systems that can do all of the key tasks humans can do to advance science and technology. If that happens, we'll see explosive progress in science and technology. The world will quickly become extremely different from how it is today.

VIII.   AJEYA COTRA
             Senior program officer at Open Philanthropy

I think my median timeline for AGI now is somewhere in the late 2030s or early 2040s, by which I mean when 99 percent of remote jobs can be done by AI systems.

IX.   DWARKESH PATEL
So, your timeline for when AI could write a Gwern-like essay is two to three years?

                     GWERN BRANWEN
               Freelance writer and researcher

I have ideas for doing that which might not require AGI. Many essay ideas are already mostly done in my corpus. You don't need to be superintelligent to finish those. For AGI in general, yes, the Anthropic timeline of 2028 seemed like a good starting point for planning purposes. Because even if you're wrong... Well, probably a lot of your writing projects weren't going to get done in the next three years anyway. It's not like you lost much by writing down only a description [of what you would have written, to prompt the AI]. You can always go back and do it yourself.

X.   DWARKESH PATEL
By 2025, what are the odds that AI kills or disempowers all of humanity?

ELIEZER YUDKOWSKY
Cofounder of the Machine Intelligence Research Institute
I have refused to deploy timelines with fancy probabilities consistently for many years. They're just not my brain's native format, and every time I try to do this, it ends up making me stupider.

DWARKESH PATEL
When you say that rounding to the nearest number, there's basically a 0 percent chance humanity survives if we keep going like this, does that include the probability of there being errors in your model?[113]

ELIEZER YUDKOWSKY
My model no doubt has many errors. The trick would be to find an error that [when corrected, means AGI alignment can actually] work. Usually, when you're trying to build a rocket, and your model of rockets is lousy, it doesn't cause the rocket to launch using half the fuel, go twice as far, and land twice as precisely on target as your calculations claimed.[114]

XI.   CARL SHULMAN
Independent adviser to Open Philanthropy
Here's a world where AI progress stalls out. You go to a $100 billion training run, but over the years progress turns out to stall. You stop getting the gains you're currently getting from moving researchers into AI from other fields. You tap out those resources, because AI is, by then, a large proportion of all research. Now you've put in all of these inputs, but they haven't yielded AGI yet. I think these inputs probably *would* yield an intelligence explosion, but if they don't, we've exhausted our current scale-up of the share of our economy trying to make AI. If that's not enough, then you have the slow grind of things like general economic growth, population growth, and so on. Progress slows down.

Therefore, the chance of advanced AI happening is relatively concentrated in the next 10 years [between 2024 and 2034] compared to the rest of the century, because we can't keep going with our current rapid redirection of resources into AI. That's a one-time thing.

Conclusion

Man is a rope stretched between the animal and the Overman—a rope over an abyss. A dangerous crossing, a dangerous wayfaring, a dangerous looking-back, a dangerous trembling and halting.
—Friedrich Nietzsche, 1883

In our interview, Dario Amodei mentioned "The Big Blob of Compute," an unpublished memo he wrote at OpenAI in 2017—before scaling laws, before the Transformer, and before LLMs. The memo asserted that if we make a neural network big enough, aim it at a broad enough distribution of data, and don't fuck it up in some other way, we'll get AGI.

Suppose that's correct, and it happens this decade. Where, exactly, is the surprise in that? The strong scaling hypothesis is holding up. Models are arriving on the timeline predicted by Hans Moravec in 1998, Ray Kurzweil in 2005, Shane Legg in 2009, and Amodei in 2018.• Skeptics' claims continue to be disproven,[115] hard benchmarks keep falling,•• and even the ARC challenge has fallen.•••

If deep learning were destined to plateau, wouldn't it have happened sooner? GPT-4 seems like an unusual point to end the story. Why would we run out of momentum now—after surpassing the Turing test, after the models have aced physics exams, and after they've taken over writing half of Google's code? If we can achieve something this general simply by investing $100 million in a

•   They weren't specific, though. The idea was "use more compute" rather than, say, "use self-attention neural networks trained autoregressively on the entire internet." Notably, Moravec does not address the data requirement.

••   The best model performance on SWE-bench, a challenging test of real-world coding ability, is currently at 22 percent, up from 2 percent in 2023. In 2024, Google DeepMind's AlphaGeometry 2 system attained the equivalent of a silver medal in the International Mathematical Olympiad, one of the most demanding cognitive competitions in the world. However, most of the heavy lifting was performed not by LLMs but by a symbolic solver. Jimenez et al., "SWE-Bench"; AlphaProof and AlphaGeometry Teams, "AI Achieves Silver-Medal Standard."

••• We actually called this in an earlier draft of this book: "One way models might surpass the ARC challenge is through inference scaling, as predicted by Leopold Aschenbrenner." Greenblatt, "Getting 50% (SoTA) on ARC-AGI"; Pfau, "Will the ARC-AGI Grand Prize Be Claimed by End of 2025?"; Greenblatt, "When Will 85% Be Reached on the Public Evaluation Set on ARC-AGI-Pub?"

broad scrape of the internet, what might happen if we increase the scale of the blob 100x? Or 10,000x?•

But then we encounter the nitty-gritty problems of scaling further. As discussed in Chapter 5, getting 10,000x effective compute into a single training run is easier said than done. If we don't reach AGI by 2040, I suspect it will be because the answer to one of the following sets of questions was no:

- Can we maintain the exponential increases in inputs required to scale training runs (and reasoning)? Will models be able to evaluate their own output and contribute to reducing their own loss? The finite supply of high-quality language tokens forms a data wall. The pole-vault over this wall will require synthetic data or reinforcement learning. In either case, models must learn how to assess themselves not only in areas with clear success criteria like games, math, and coding, but also in more ambiguous domains. (Where is the RL environment that teaches you to beat Lyndon Johnson at political 4D chess?) Also, will labs crack multi-data-center training? It might not be feasible to get more than 1 GW into a single data center. But if they can train a model across multiple sites with acceptable performance penalties, then there may still be hope for scaling beyond 1 GW of GPUs.

- Will LLMs ever demonstrate the kind of generalization we see in humans? Compared to humans, these models are remarkably inefficient at learning. We train them on data it would take a person 10,000 years to accumulate. Their learning process seems incompatible with the Einstein-level creativity we see in humans—the flash of insight derived from thought experiments and

---

• There's no need to overfocus on current LLMs or the Transformer neural network. Consider two hypotheses that may be even more likely than the LLM scaling hypothesis, both of which suggest that LLMs could give rise to successor systems leading to AGI:

1. The bootstrap scaling hypothesis: The idea that LLMs will help AI researchers design a replacement system, and that this system or its successor will achieve human-level intelligence.
2. The profit scaling hypothesis: The idea that revenue from LLM-based products will fund sufficient AI research to produce a non-LLM AGI.

a handful of contradictory observations.• Will this limitation diminish with larger models and improved training techniques? If not, can the models' vast crystallized intelligence compensate for their lack of fluid intelligence? Is fluid intelligence in fact just a recombination of crystallized intelligence?

- Will future algorithmic improvements stack onto one another? Can they continue to substitute for actual compute? If not, the fast takeoff scenarios, in which a sufficiently advanced model bootstraps itself to super-intelligence by discovering compute efficiency gains, thereby enabling you to quickly train a successor model, become implausible.

In my estimation, each of these challenges is quite likely to be solved. However, *all* of them must be addressed to scale to AGI. The need to consider the cumulative probability of success lowers my estimated probability of AGI by 2040 down to 60 percent.

Over the past two years, I've spoken with lab leaders, scientists, philosophers, and forecasters, all at the top of their game. They disagree a lot. Those closest to the answers are, as you may have noticed, often evasive. Being too close to the proverbial elephant may instill biases that make their overall picture less reliable. It's unclear whether these biases net out as overly optimistic or overly pessimistic. Ilya Sutskever told me that the researchers closest to the frontier are often the most pessimistic, because transitional problems—bugs that are trivial in the grand scheme of things—are especially salient in their daily work, while the yearslong trends tend to fade into the background.

One frustrating aspect of day-to-day AI news coverage is its focus on interpersonal drama or rumors of some supposed secret method for advancing AI capabilities. The "What did Ilya see?" meme from 2023 captured both of these tendencies.•• Of course, it does matter who leads the frontier labs and what their priorities are; history shows

---

• For example, one of the few pieces of direct evidence available to Einstein when he was developing general relativity was an anomaly in Mercury's orbit. However, he also had substantial indirect data, mathematical insights, and a robust theoretical tradition to support his work.

•• A 2023 meme capturing widespread confusion about why Sutskever joined the effort to oust Sam Altman from OpenAI. See Schroeder, "What Did He See?"

that such personnel issues can be critically important.• But the real story of AI is the consistency and outlandish success of the trend line. In just over a decade, we've progressed from neural networks that could perform only one narrow task, such as identifying one of 1,000 prespecified species in a blurry photo, to AIs capable of answering an extraordinary range of real-world medical, legal, and programming questions.

While it is unsettling to consider how wrong things could go, I believe AI is more likely than not to be net beneficial for humanity.•• The whimsical pronouncements of AI developers are, in fact, valid: AI really could help us alleviate life's hardships, lessen the misery of disease, and restore the beauty and connection we lose due to scarcity.[116]

Moreover, there's no going back. It seems that the universe is structured such that throwing large amounts of compute at the right distribution of data gets you AI. And the secret is out. If the scaling picture is roughly correct, it's hard to imagine AGI *not* being developed this century, even if some actors hold back, or are held back.

Despite the book's title and subtitle, I'm not writing about the settled past. The purpose of this book is to capture what it feels like to be in the midst of the scaling era. Naturally, settled answers are unavailable; we're all running unsupervised. So I've done my best simply to ask the right questions. I'm incredibly excited to keep asking questions—better questions. The answers will come in their own time.

Goodness, it's going to be so interesting.

•   Consider Henry Kissinger reportedly preventing Richard Nixon from launching a nuclear strike on North Korea, or Vasili Arkhipov keeping his submarines from launching nuclear missiles at the US. Lee, "Did Richard Nixon Order Nuclear Strike on North Korea While Drunk?"; Walsh, "This Man Stopped the Cuban Missile Crisis from Going Nuclear."

••   One large survey of AI researchers provides ambiguous support for this claim. According to the authors, "68.3 percent thought good outcomes from superhuman AI are more likely than bad." Grace et al., "AI Authors on the Future of AI."

Profiles

DARIO AMODEI

Cofounder and CEO of Anthropic. Previously, he served as vice president of research and led the safety team at OpenAI. He coauthored the GPT-3 paper, "Language Models Are Few-Shot Learners," which demonstrated that a sufficiently large language model can learn on the fly.[117] He was also the lead author of the seminal paper "Concrete Problems in AI Safety," which was the first to codify the empirical side of the field. Additionally, he led work on Deep Speech 2, an early end-to-end speech recognition system, at Baidu.[118]

LEOPOLD ASCHENBRENNER

Cofounder of Situational Awareness LP, an investment firm focused on AGI. Previously, he worked on the super-alignment team at OpenAI, with the FTX Future Fund, and as a research fellow at the Global Priorities Institute at Oxford. His book-length essay "Situational Awareness," which forecasts the near-term future of AI, received widespread attention in policy circles and influenced expectations around state actors intervening in the development of AGI.[119]

GWERN BRANWEN

Pseudonymous researcher well known to insiders for his incisive writing on AI, metascience, internet culture, and rationality. His work often anticipates the technical details of AI. In 2020, he coined the term "scaling hypothesis" to describe the strategy adopted by AI labs in subsequent years.[120] Aside from his thousands of deeply researched comments on public forums, his most notable work is probably his website, gwern.net, which includes a vast collection of essays, aphorisms, and personal archives.

TRENTON BRICKEN

Researcher on Anthropic's mechanistic interpretability team. He has worked on attention mechanisms and the utility of sparse representations in neural networks.[121] He also coauthored the groundbreaking paper "Towards Monosemanticity," a technique for interpreting the concepts large models use.[122] Previously, he was a PhD student at Harvard, where he studied the relationships between machine learning and neuroscience.

JOE CARLSMITH

Senior research analyst at Open Philanthropy. His work focuses on existential risks from AI, the complexity of human values, and potential futures. He completed his

PhD in philosophy at Oxford in 2023, covering technical topics in epistemology and ethics, including anthropic reasoning, simulation arguments, and infinite ethics. He is best known for his model of power-seeking AI and its probability.[123] He regularly publishes on joecarlsmith.com, and is currently working on a book-length essay series titled "Otherness and Control in the Age of AGI."[124]

FRANÇOIS CHOLLET

Senior staff software engineer at Google and creator of the Keras deep learning library and the Abstraction and Reasoning Corpus (ARC-AGI), a set of puzzles designed to test for general intelligence, a benchmark that current leading models still struggle with. He is the author of the book *Deep Learning With Python* and has led significant research on deep learning for computer vision, including the Xception architecture.[125]

AJEYA COTRA

Senior program officer for potential risks from advanced AI at Open Philanthropy, and its primary grantmaker for technical AI safety research. Best known for her model of AI timelines, which estimates when an AI training run will use the same amount of compute as the evolution of the human brain did.[126] She blogs at planned-obsolescence.org.

TYLER COWEN

Professor of economics at George Mason University and chairman of the Mercatus Center, a libertarian think tank. He coauthors the blog Marginal Revolution and hosts the *Conversations with Tyler* podcast. He also administers the Emergent Ventures grants program, which aims to identify unconventional early-career talent. In 2023, he released a generative book, *GOAT: Who Is the Greatest Economist of All Time and Why Does It Matter?,* a free GPT-4 fine-tune allowing readers to talk to the text and remix it.

SHOLTO DOUGLAS

Reinforcement learning infrastructure lead at Anthropic. Previously, he was a machine learning researcher at Google DeepMind, where he led the Gemini inference and reasoning teams and worked on open-source models and improving the efficiency of Transformer models.[127] Prior to joining Google DeepMind, he was a consultant at McKinsey & Company and spent a year at the Australian Centre for Field Robotics.

DEMIS HASSABIS

Cofounder and CEO of Google DeepMind, an AI startup acquired by Google in 2014, and one of the first companies to set the creation of AGI as an explicit goal. During his PhD at University College London, he researched the implementation of human episodic memory. He played a key role in several of DeepMind's major breakthroughs, including AlphaGo, AlphaZero, and AlphaFold, which demonstrated deep learning's ability to excel at tasks once believed to be too complex and nuanced for computers to perform.[128] In 2024, he was awarded a UK knighthood and the Nobel Prize in Chemistry.

JARED KAPLAN

Cofounder of Anthropic and associate professor of physics at John Hopkins University. Previously, he served as a research consultant at OpenAI. His research spans theoretical physics and machine learning, with significant contributions to our understanding of neural networks and the application of LLMs for coding.[129] He led work on "Scaling Laws for Neural Language Models," a seminal paper describing how the performance of neural language models improves with increasing resources.[130] He has also contributed to research on RLHF and red-teaming LLMs.[131]

HOLDEN KARNOFSKY

Member of technical staff at Anthropic and former CEO and director of AI strategy at Open Philanthropy. His writing has significantly influenced the Effective Altruism movement, of which he is a vocal supporter. Two of his most influential pieces are "Hits-Based Giving," an argument for backing philanthropic ideas with the explicit risk of failure, and "The Most Important Century," a book-length argument for the urgency of AI safety.[132]

SHANE LEGG

Cofounder and chief AGI scientist at Google DeepMind. His PhD research focused on the theory of general intelligence. In 2007, he coauthored an influential formalization of universal intelligence.[133] He is also credited with reintroducing the term "AGI" in 2002, at a time when the possibility of general AI was met with deep skepticism in the scientific community.[134] He has contributed to significant advancements in deep reinforcement learning, including the Deep-Q Network, a breakthrough that enhanced AI's ability to play video games at human level using raw data, which revolutionized reinforcement learning.[135]

DYLAN PATEL

Chief analyst at SemiAnalysis, a research and consulting firm specializing in the global semiconductor supply chain. Its model of the data center industry is notable for using satellite imagery to identify lesser-known construction projects. Previously, he spent years as an independent researcher and moderated r/hardware, one of Reddit's largest subreddits, with 4.1 million members as of November 2024. His essay with Daniel Nishball, "Google Gemini Eats the World," is notable for coining the term "GPU-rich."[136]

JOHN SCHULMAN

Senior researcher at Anthropic. He is a cofounder of OpenAI and previously led their post-training team, including the reinforcement learning work that turned GPT-3 into ChatGPT and WebGPT, one of the first tool-augmented language models.[137] He is the lead author on two of the most influential reinforcement learning papers of all time: his PhD thesis on trust-region policy optimization and "Proximal Policy Optimization Algorithms," which simplified and stabilized the training of neural policies.[138]

CARL SHULMAN

Independent adviser to Open Philanthropy on technological progress and risk. From 2012 to 2024, he was a research associate at the University of Oxford's Future of Humanity Institute. Previously, he served as director of careers research at the charity 80,000 Hours and modeled intelligence explosion scenarios at the Machine Intelligence Research Institute (MIRI). He has introduced concepts and models across a range of fields, including genetics, decision theory, cause prioritization, and economics. He blogs at reflectivedisequilibrium.blogspot.com.

ILYA SUTSKEVER

Cofounder and chief scientist of Safe Superintelligence Inc., a new frontier AI lab. He is also a cofounder of OpenAI and served as its chief scientist from 2015 to 2024. As a student in Geoffrey Hinton's lab at the University of Toronto, he coauthored the AlexNet paper, a breakthrough entry to the ImageNet contest that demonstrated the power of deep learning.[139] This led to the resurgence of neural networks and is often regarded as the beginning of the deep learning revolution. He is also known for his work on DeepMind's AlphaGo system and on Seq2Seq, research that transformed machine translation and paved the way for large language models.[140] He was elected a Fellow of the Royal Society in 2022.

JON Y
Creator of the YouTube channel Asianometry, known for its original content on the history and current state of semiconductor manufacturing. He also runs the Asianometry newsletter. He is based in Taipei.

ELIEZER YUDKOWSKY
Cofounder of the influential AI safety lab the Machine Intelligence Research Institute, which he helped establish in 2000 at the age of 21. He is also a well-known writer of fiction and essays on science and rationality. His work, often published on LessWrong, the forum he founded, routinely generates intense debate. Notable for arguing that AI alignment cannot be achieved in time to prevent harm from AGI, he is perhaps best known for his 2023 op-ed in *Time* advocating for a state-backed global halt to AI development, a piece that significantly advanced public awareness of AI safety concerns.[141]

MARK ZUCKERBERG
Chairman and CEO of Meta Platforms, the parent company of Facebook, Instagram, and WhatsApp, which together form one of the world's most influential social media ecosystems. In AI, he is notable for advocating for open model weights, a stance that has led Meta's AI research team to release some of the company's most advanced models.[142]

Appendix: Essays

"The Bitter Lesson"
Richard Sutton

March 2019

The biggest lesson that can be read from 70 years of AI research is that general methods that leverage computation are ultimately the most effective, and by a large margin. The ultimate reason for this is Moore's law, or rather its generalization of continued exponentially falling cost per unit of computation. Most AI research has been conducted as if the computation available to the agent were constant (in which case, leveraging human knowledge would be one of the only ways to improve performance), but over a slightly longer time than a typical research project, massively more computation inevitably becomes available. Seeking an improvement that makes a difference in the shorter term, researchers seek to leverage their human knowledge of the domain, but the only thing that matters in the long run is the leveraging of computation.

These two need not run counter to each other, but in practice they tend to. Time spent on one is time not spent on the other. There are psychological commitments to investment in one approach or the other. And the human knowledge approach tends to complicate methods in ways that make them less suited to taking advantage of general methods leveraging computation. There were many examples of AI researchers' belated learning of this bitter lesson, and it is instructive to review some of the most prominent.

In computer chess, the methods that defeated the world champion, Kasparov, in 1997 were based on massive, deep search. At the time, this was looked upon with dismay by the majority of computer chess researchers, who had pursued methods that leveraged human understanding of the special structure of chess. When a simpler search-based approach with special hardware and software proved vastly more effective, these human-knowledge-based chess researchers were not good losers. They said that "brute force" search may have won this time, but it was not a general strategy, and anyway it was not how people played chess. These researchers wanted methods based on human input to win and were disappointed when they did not.

A similar pattern of research progress was seen in computer Go, only delayed by a further 20 years. Enormous initial efforts went into avoiding search by taking advantage of human knowledge or the special features of the game, but all those efforts proved irrelevant, or worse, once search was applied effectively at scale. Also important was the use of learning by self-play to learn a value function (as it was in many other games, and even in chess, although learning did not play a big role in the 1997 program that first beat a world champion). Learning by self-play, and learning in general, is like search in that it enables massive computation to be brought to bear. Search and learning are the two most important classes of techniques for utilizing massive amounts of computation in AI research. In computer Go, as in computer chess, researchers' initial effort was directed toward utilizing human understanding (so that less search was needed), and only much later was much greater success had by embracing search and learning.

In speech recognition, there was an early competition, sponsored by DARPA, in the 1970s. Entrants included a host of special methods that took advantage of human knowledge: knowledge of words, of phonemes, of the human vocal tract, et cetera. On the other side were newer methods that were more statistical in nature and did much more computation, based on hidden Markov models (HMMs). Again, the statistical methods won out over the human-knowledge-based methods. This led to a major change in all of natural language processing, gradually over decades, where statistics and computation came to dominate the field. The recent rise of deep learning in speech recognition is the most recent step in this consistent direction. Deep learning methods rely even less on human knowledge and use even more computation, together with learning on huge training sets, to produce dramatically better speech recognition systems. As in the games, researchers always tried to make systems that worked the way the researchers thought their own minds worked—they tried to put that knowledge in their systems—but it proved ultimately counterproductive, and a colossal waste of researchers' time, when, through Moore's law, massive computation became available and a means was found to put it to good use.

In computer vision, there has been a similar pattern. Early methods conceived of vision as searching for edges, or generalized cylinders, or in terms of SIFT features. But today all this is discarded. Modern deep-learning neural networks use only the notions of convolution and certain kinds of invariances, and perform much better.

This is a big lesson. As a field, we still have not thoroughly learned it, as we are continuing to make the same kinds of mistakes. To see this, and to effectively resist it, we have to understand the appeal of these mistakes. We have to learn the bitter lesson that building in how we think we think does not work in the long run. The bitter lesson is based on the historical observations that 1) AI researchers have often tried to build knowledge into their agents; 2) this always helps in the short term, and is personally satisfying to the researcher; but 3) in the long run it plateaus and even inhibits further progress; and 4) breakthrough progress eventually arrives by an opposing approach based on scaling computation by search and learning. The eventual success is tinged with bitterness, and often incompletely digested, because it is success over a favored, human-centric approach.

One thing that should be learned from the bitter lesson is the great power of general-purpose methods, of methods that continue to scale with increased computation even as the available computation becomes very great. The two methods that seem to scale arbitrarily in this way are *search* and *learning*.

The second general point to be learned from the bitter lesson is that the actual contents of minds are tremendously, irredeemably complex; we should stop trying to find simple ways to think about the contents of minds, such as simple ways to think about space, objects, multiple agents, or symmetries. All of these are part of the arbitrary, intrinsically complex, outside world. They are not what should be built in, as their complexity is endless; instead, we should build in only the meta-methods that can find and capture this arbitrary complexity. Essential to these methods is that they can find good approximations, but the search for them should be by our methods, not by us. We want AI agents that can discover like we can, not ones that contain what we have discovered. Building in our discoveries only makes it harder to see how the discovering process can be done.

Excerpt from "The Scaling Hypothesis"
Gwern Branwen

May 2020

The pretraining thesis goes something like this:

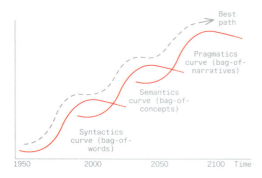

Figure 1. Envisioned evolution of NLP research through three different eras or curves: the hypothetical S-curves and progress in natural language modeling. Cambria and White, "Jumping NLP Curves."

Humans, one might say, are the cyanobacteria of AI. We constantly emit large amounts of structured data, which implicitly rely on logic, causality, object permanence, history—all that good stuff. All of that is implicit and encoded into our writings and videos and "data exhaust." A model learning to predict must learn to understand all that to get the best performance; as it predicts the easy things, which are mere statistical pattern matching, what's left are the hard things. AI critics often say that the long tail of scenarios for tasks like self-driving cars or natural language can only be solved by true generalization and reasoning. It follows, then, that if models solve the long tail, they must learn to generalize and reason.

Early on in training, a model learns the crudest levels: that some letters, like E, are more frequent than others, like Z; that every five characters or so there is a space; and so on. It goes from predicted, uniformly distributed bytes to what looks like Base60 encoding—alphanumeric gibberish. As crude as this may be, it's enough to make quite a bit of absolute progress. A random predictor needs 8 bits to "predict" a byte or character, but just by matching letter and space frequencies, it can almost halve its error to around 5 bits. Because it's learning so much from every character, and because the learned frequencies

are simple, it can happen so fast that if one is not logging samples frequently, one might not even observe the improvement.

As training progresses, the task becomes more difficult. Now the model begins to learn which words actually exist and do not exist. It doesn't know anything about meaning, but at least now when it's asked to predict the second half of a word, it can actually do that to some degree, saving it a few more bits. This takes a while, because any specific instance will show up only occasionally. A word may not appear in a dozen samples, and there are many thousands of words to learn. With some more work, it has learned that punctuation, pluralization, and possessives are all things that exist. Put that together, and it may have progressed again, all the way down to 3 to 4 bits error per character! (While the progress is gratifyingly fast, it's still all gibberish, make no mistake. A sample may be spelled correctly but doesn't make even a bit of sense.)

But once a model has learned a good English vocabulary and correct formatting and spelling, what's next? There's not much juice left in predicting within words. The next thing is to pick up on associations among words. Which words tend to come first? Which words cluster and are often used near each other? Nautical terms tend to get used a lot with each other in sea stories. Ditto words in Bible passages, or American history Wikipedia articles, and so on. If the word "Jefferson" is the last word, then "Washington" may not be far away, so the model should hedge its bets on predicting that W is the next character, and then, if it shows up, go all in on "ashington." Such bag-of-words approaches still predict badly, but now we're down to perhaps less than 3 bits error per character.

What next? Does it stop there? Not if there's enough data and the earlier stuff, like learning English vocabulary, doesn't hem the model in by using up its learning ability. Gradually, other words, like "president" or "general" or "after," begin to show the model subtle correlations: "Jefferson was president after..." With many such passages, the word "after" begins to serve a use in predicting the next word, and then the use can be broadened.

By this point, the loss is perhaps 2 bits;

every additional 0.1-bit decrease comes at a steeper cost and takes more time. However, now the sentences have started to make sense. A sentence like "Jefferson was president after Washington" does in fact mean something (and if, occasionally, we sample "Washington was president after Jefferson," well, what do you expect from such an unconverged model?). Jarring errors will immediately jostle us out of any illusion about the model's understanding, and so training continues. (Around here, Markov chain and n-gram models start to fall behind; they can memorize increasingly large chunks of the training corpus, but they can't solve increasingly critical syntactic tasks like balancing parentheses or quotes, much less start to ascend from syntax to semantics.)

Now training is hard. Even subtler aspects of language must be modeled, such as keeping pronouns consistent. This is hard in part because the model's errors are becoming rare, and in part because the relevant pieces of text are increasingly distant and long range. As it makes progress, the absolute size of errors shrinks dramatically. Consider the case of associating names with gender pronouns. The difference between "Janelle ate some ice cream because he likes sweet things" and "Janelle ate some ice cream because she likes sweet things" is one no human could fail to notice, yet it is a difference of a single letter. If we compared two models, one that didn't understand gender pronouns at all and guessed "he" or "she" purely at random, and one that understood them perfectly and always guessed "she," the second model would attain a lower average error of barely <0.02 bits per character!

Nevertheless, as training continues, these problems and more, like imitating genres, get solved, and eventually at a loss of 1 to 2 bits (where a small char-RNN might converge

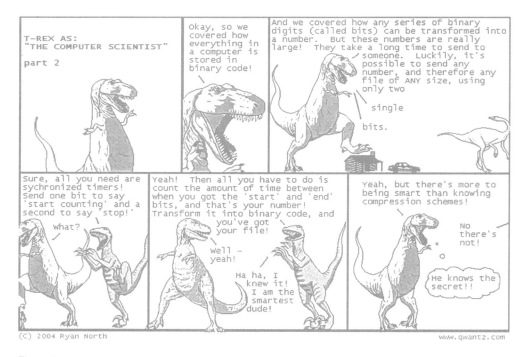

Figure 2.
"Yeah, but there's more to being smart than knowing compression schemes!"
"No there's not!"
"He knows the secret!"
—Ryan North, "T-Rex as: 'The Computer Scientist,' Part 2"

on a small corpus like Shakespeare or some Project Gutenberg e-books), we will finally get samples that sound human—at least for a few sentences. These final samples may convince us briefly, but, aside from issues like repetition loops, even with good samples, the errors accumulate. A sample will state that someone is "alive" and then 10 sentences later use the word "dead," or it will digress into an irrelevant argument instead of the expected next argument, or someone will do something physically improbable, or it may just continue for a while without seeming to get anywhere.

All of these errors are far less than <0.02 bits per character. We are now talking not hundredths of bits per character but less than ten-thousandths.

The pretraining thesis argues that this can go even further: We can compare this performance directly with humans doing the same objective task, who can achieve closer to 0.7 bits per character. What is in that missing >0.4?

Well, *everything*! Everything that the model misses. While just babbling random words was good enough at the beginning, at the end it needs to be able to reason its way through the most difficult textual scenarios requiring causality or common-sense reasoning. Every error where the model predicts that ice cream put in a freezer will "melt" rather than "freeze"; every case where the model can't keep straight whether a person is alive or dead; every time the model chooses a word that doesn't help build toward the ultimate conclusion of an essay; every time it lacks the theory of mind to compress novel scenes describing the Machiavellian scheming of a dozen individuals at dinner jockeying for power as they talk; every use of logic or abstraction or instruction or Q&A where the model is befuddled and needs more bits to cover up for its mistake, where a human would think, understand, and predict. For a language model, the truth is that which keeps on predicting well, because truth is one and error many. Each of these cognitive breakthroughs allows for ever so slightly better prediction of a few relevant texts; nothing less than true understanding will suffice for ideal prediction.

If we trained a model that reached that loss of <0.7, that could predict text indistinguishably from a human, whether in a dialogue or when quizzed about ice cream or tested on SAT analogies or tutored in mathematics; if, for every string, the model did just as good a job of predicting the next character as you could do, then how could we say that it doesn't truly understand everything? (If nothing else, we could, by definition, replace humans in any kind of text-writing job!)

The last bits are deepest. The implication here is that the final few bits are the most valuable bits, which require the most of what we think of as intelligence. A helpful analogy here might be our own actions. For the most part, all humans execute actions equally well. We can all pick up a tea mug without dropping it and lift our legs to walk down thousands of steps without falling even once. For everyday actions (the sort that make up most of a corpus), anybody, of any intelligence, can get enough practice and feedback to do them quite well, learning individual algorithms to solve each class of problem extremely well in isolation.[1] Meanwhile, for rare problems, there may be too few instances to do any better than memorize the answer. In the middle of the spectrum are problems that are similar but not too similar to other problems; these are the sorts of problems that reward flexible meta-learning and generalization, and many intermediate problems may be necessary to elicit those capabilities ("Neural nets are lazy").

Where individuals differ is when they start running into the long tail of novel choices, rare choices, choices that take seconds but unfold over a lifetime, choices where we will never get any feedback (like after our death). One only has to make a single bad decision out of a lifetime of millions of discrete decisions to wind up in jail or dead. A small absolute average improvement in decision quality, if it pertains to those decisions, may be far more important than their quantity indicates, which gives us some intuition for why those last bits are the hardest and deepest. (Why do humans have such large brains, when animals like chimpanzees do so many ordinary activities seemingly as well with a fraction of the expense? Why is language worthwhile?

Perhaps because of considerations like these. We may be at our most human when filling out the paperwork for life insurance.)

Reasons for doubt. The pretraining thesis, while logically impeccable—how is a model supposed to solve all possible trick questions without understanding, just guessing?—never struck me as convincing, an argument admitting neither confutation nor conviction. It feels too much like a magic trick: "Here's some information theory, here's a human benchmark, here's how we can encode all tasks as a sequence prediction problem. Hey presto—intelligence!" There are lots of algorithms that are Turing complete or "universal" in some sense; there are lots of algorithms like AIXI that solve AI in some theoretical sense (Schmidhuber and company have many of these cute algorithms, such as "the fastest possible algorithm for all problems," with the minor catch of some constant factors that require computers bigger than the universe).

Why think pretraining or sequence modeling is not another one of them? Sure, if the model got a low enough loss, it'd have to be intelligent, but how could you prove that would happen in practice? (Training char-RNNs was fun, but they didn't exactly revolutionize deep learning.) It might require more text than exists, countless petabytes of data for all of those subtle factors, like logical reasoning, to represent enough training signal, amid all the noise and distractors, to train a model. Or maybe your models are too small to do more than absorb the simple surface-level signals, and you would have to scale them 100 orders of magnitude for it to work, because the scaling curves didn't cooperate. Or maybe your models are fundamentally broken, and stuff like abstraction requires an entirely different architecture to work at all, and whatever you do, your current models will saturate at poor performance. Or it'll train, but it'll spend all its time trying to improve the surface-level modeling, absorbing more and more literal data and facts without ever ascending to the higher planes of cognition as planned. Or...

The possibilities of developing an atomic weapon and the desirability of doing it secretly were discussed at a Princeton University conference in which I participated in March 1939... Bohr said this rare variety could not be separated from common uranium except by turning the country into a gigantic factory. Bohr was worried that this could be done and that an atomic bomb could be developed—but he hoped that neither could be accomplished. Years later, when Bohr came to Los Alamos, I was prepared to say, "You see..." But before I could open my mouth, he said, "You see, I told you it couldn't be done without turning the whole country into a factory. You have done just that." —Edward Teller[2]

But apparently it would've worked fine. Even RNNs probably would've worked. Transformers are nice, but they seem mostly to be about efficiency.[3] (Training large RNNs is much more expensive, and doing BPTT over multiple nodes is much harder engineering-wise.) It just required more compute and data than anyone was willing to risk on it until a few true believers were able to get their hands on a few million dollars of compute.

Q: Did anyone predict, quantitatively, that this would happen where it did?

A: Not that I know of.

Q: What would future scaled-up models learn?

GPT-2 1.5B had a cross-entropy WebText validation loss of ~3.3, based on the perplexity of ~10 and log2(10) = 3.32. GPT-3 halved that loss to ~1.73, judging from Brown et al. 2020, using the scaling formula $2.57 \times (3.64 \times 10^3) - 0.048$. For a hypothetical GPT-4, if the scaling curve continues for another three orders or so of compute (100–1,000x) before crossing over and hitting harder diminishing returns, the cross-entropy loss will drop to ~1.24 ($2.57 \times (3.64 \times (10^3 \times 10^3)) - 0.048$).

If GPT-3 gained so much meta-learning and world knowledge by dropping its absolute loss ~50 percent when starting from GPT-2's level, what capabilities would another ~30 percent improvement over GPT-3 gain? (Cutting the loss that much would still not reach human level, as far as I can tell.)[4] What would a drop to ≤1, perhaps using wider context windows or recurrency, gain?

A: I don't know.

Q: Does anyone?

A: Not that I know of.[5]

NOTES

1   If you see thousands of images labeled "dog" and thousands more labeled "cat," you can simply learn to separate dog and cat classifiers without bothering to understand their shared aspects, like being domesticated quadruped mammal predators. This won't be useful if you are then asked to classify "ferret" images, but you weren't asked to, so that's not your problem, since you can just learn yet another separate classifier for ferrets if you then get a lot of ferret images.

2   Edward Teller and Allen Brown, *The Legacy of Hiroshima* (Macmillan & Co), 210–11.

3   Another way of interpreting the various papers about how Transformers are actually like RNNs or are actually Hopfield networks is to take that as indicating that what is important about them is not any inherent new capability compared to older architectures but some lower-level aspect, like being more efficiently trainable on contemporary hardware.

4   How do these absolute prediction performances compare to humans? It's hard to say. The only available benchmarks for perplexity for humans, GPT-2, and GPT-3 appear to be WebText, Penn Tree Bank (PTB, based on the Brown Corpus), 1 Billion Word (1BW), and LAMBADA. But coverage is spotty. I found no human benchmarks for WebText or PTB, so I can't compare the human versus GPT-2 or GPT-3 perplexities (GPT-2 PTB: 35.7; GPT-3 PTB: 20.5).

GPT-2 was benchmarked at 43 perplexity on the 1BW benchmark versus a (highly extrapolated) human perplexity of 12 (which, interestingly, extrapolates, using 2012 LSTM RNNs, that "10 to 20 more years of research [is needed] before human performance is reached"), but that may be an unfair benchmark ("Our model is still substantially worse than prior work on the 1BW benchmark… This is likely due to a combination of it being both the largest dataset and having some of the most destructive pre-processing—1BW's

sentence-level shuffling removes all long-range structure"), and 1BW was dropped from the GPT-3 evaluation due to data contamination ("We omit the 4 Wikipedia-related tasks in that work because they are entirely contained in our training data, and we also omit the 1BW benchmark due to a high fraction of the dataset being contained in our training set").

LAMBADA was benchmarked at a GPT-2 perplexity of 8.6 and a GPT-3 perplexity of 3 (zero-shot) and 1.92 (few-shot). OpenAI claims in their GPT-2 blog post (but not the paper) that human perplexity is 1 to 2, but it provides no sources, and I couldn't find any. (The authors might be guessing based on how LAMBADA was constructed. Examples were filtered by whether two independent human raters provided the same right answer, which lower-bounds how good humans must be at predicting the answer.)

So overall, it looks like the best guess is that GPT-3 continues to have somewhere around twice the absolute error of a human. This implies it will take a large (yet far from impossible) amount of compute to fully close the remaining gap with the current scaling laws. If we irresponsibly extrapolate out the WebText scaling curve further, assume GPT-3 has twice the error of a human at its current WebText perplexity of 1.73 (and so humans are ~0.86), then we need $2.57 \cdot (3.64 \cdot (103 \cdot x)) - 0.048 = 0.86$, where $x = 2.2^6$, or 2,200,000x the compute of GPT-3. (This would roughly equal the cost of the US invading Iraq.)

When is that feasible? If we imagine that peak AI compute usage doubles every 3.4 months, then 2.2^6 would be 22 doublings away, or 6.3 years, in 2027. Most people believe that compute trend must break down soon, and that sort of prediction is a good reason why!

Going the other direction, Hernandez and Brown 2020's estimate is that, net of hardware and algorithmic progress, the cost of a fixed level of performance halves every 16 months. So if GPT-3 cost ~$5 million in early 2020, then it'll cost $2.5 million around mid-2021, and so on. Similarly, a GPT-human requiring 2.2^6x more compute would presumably cost on the order of $10 trillion in 2020, but after 14 halvings (18 years) would cost $1 billion in 2038.

5   As of December 2020, half a year later, almost no researcher has been willing to go on record as saying what specific capabilities they predict future 1 trillion, 10 trillion, or 100 trillion parameter models will have or not have, and at what size which missing capabilities will emerge—just as no one is on record successfully predicting GPT-2's or GPT-3's specific capabilities.

"Will Scaling Work?"
Dwarkesh Patel

December 2023

When should we expect AGI?

If we can keep scaling LLMs++ and get better and more general performance as a result, then there's reason to expect powerful AIs by 2040 (or much sooner) that can automate most cognitive labor and speed up further AI progress. However, if scaling doesn't work, then the path to AGI seems much longer and more intractable, for reasons I explain in this piece.

In order to think through both the pro and con arguments about scaling, I've written it as a debate between two characters I made up: Believer and Skeptic.

Will we run out of data?

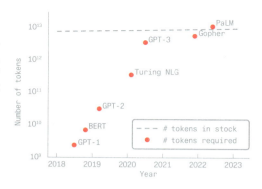

Figure 1. Insufficient tokens hinder LLM scaling. Villalobos et al., "Will We Run Out of Data?"

SKEPTIC

We're about to run out of high-quality language data next year.[1]

Even taking handwavy scaling curves seriously implies that we'll need 10^{35} FLOPs for an AI that is reliable and smart enough to write a scientific paper.[2] That's table stakes for the abilities an AI would need to automate further AI research and continue to progress once scaling becomes infeasible.[3] Which means we need 5 OOMs more data than we seem to have.[4]

I'm worried that when people hear "We're 5 OOMs off," how they register it is "We have 5x less data than we need—we just need a couple of 2x improvements in data efficiency, and we're golden." After all, what's a couple OOMs between friends?

No, 5 OOMs off means we have 100,000x less data than we need. Yes, we'll get slightly more data-efficient algorithms, and multimodal training will give us more data, plus we can recycle tokens on multiple epochs and use curriculum learning.[5] But even if we assume the most generous possible one-off improvements from these techniques, they do not give us the exponential increase in data we need to keep up with the exponential increase in compute demanded by these scaling laws.

So then people say we'll get self-play and synthetic data working somehow. But self-play has two very difficult challenges:

— Evaluation: Self-play worked with AlphaGo

because the model could judge itself based on a concrete win condition: Did I win this game of Go? But novel reasoning doesn't have a concrete win condition. As a result, just as you would expect, LLMs are so far incapable of correcting their own reasoning.[6]

— Compute: All these math and code approaches tend to use various sorts of tree search, where you run an LLM at each node repeatedly. AlphaGo's compute budget is staggering for the relatively circumscribed task of winning at Go.[7] Now imagine that instead of searching over the space of Go moves, you need to search over the space of all possible human thought. All this extra compute needed to get self-play to work is in addition to the stupendous compute increase already required to scale the parameters themselves (compute = parameters × data). Using the 1^{35} FLOP estimate for human-level thought, we need 9 OOMs more compute atop the biggest models we have today.[8] Yes, you'll get improvements from better hardware and better algorithms. But will you really get a full equivalent of 9 OOMs?

BELIEVER

If your main objection to scale working is just the lack of data, your intuitive reaction should not be "Well, it looks like we could have produced AGI by scaling up a Transformer++, but I guess we're gonna run out of data first." Your reaction should be "Holy fuck—if the internet was a lot bigger, scaling up a model whose basic structure I could write in a few hundred lines of Python code would have produced a human-level mind."[9] It's a crazy fact about the world that it's this easy to make big blobs of compute intelligent.

The sample over which LLMs are "inefficient" is mostly just irrelevant e-commerce junk.[10] We compound this disability by training them on predicting the next token, a loss function that is almost completely unrelated to the actual tasks we want intelligent agents to do in the economy. And despite the minuscule intersection between the abilities we actually want and the terrible loss function and data we train these models with, we can produce a baby AGI (aka GPT-4) by throwing just 0.03 percent of Microsoft's yearly revenues at a big scrape of the internet.[11]

So, given how easy and simple AI progress has been so far, we shouldn't be surprised if synthetic data also just works. After all, "the models just want to learn."[12]

GPT-4 has been out for all of eight months. The other AI labs have only just gotten their own GPT-4-level models. Which means that all the researchers are only now getting around to making self-play work with current-generation models (and it seems like one of them might have already succeeded).[13] Therefore, the fact that so far we don't have public evidence that synthetic data has worked at scale doesn't mean it can't.

After all, RL becomes much more feasible when your base model is capable enough to get the right answer at least some of the time. Now you can reward that 1 out of 100 times the model accomplishes the chain of thought required for an extended math proof, or writes the 500 lines of code needed to complete a full pull request. Soon your 1 in 100 success rate becomes 10 in 100, then 90 in 100. Now you try the 1,000-line pull request, and not only will the model sometimes succeed, but it'll also be able to critique itself when it fails. And so on.

In fact, this synthetic data bootstrapping seems almost directly analogous to human evolution. Our primate ancestors show little evidence of being able to rapidly discern and apply new insights.[14] But once humans develop language, you have this genetic and cultural coevolution that is very similar to the synthetic data–self-play loop for LLMs, where the model gets smarter in order to better make sense of the complex symbolic outputs of similar copies.[15]

Self-play doesn't require models to be perfect at judging their own reasoning. They just have to be better at evaluating reasoning than at doing it de novo (which clearly already seems to be the case—see Constitutional AI, or just play around with GPT for a few minutes, and notice that it seems to be better at explaining why what you wrote down is wrong than it is at coming up with the right answer by itself).[16]

Almost all the researchers I talk to in the big AI labs are quite confident they'll get self-play to work. When I ask why they're so sure, they heave for a moment, as if they're bursting to explain all their ideas. But then they remember that confidentiality is a thing and say, "I can't tell you specifics, but there's so much low-hanging fruit in terms of what we can try here." Or, as Dario Amodei, CEO of Anthropic, told me on my podcast:

Dwarkesh Patel
You mentioned that data is likely not to be the constraint. Why do you think that's the case?

Dario Amodei
There are various possibilities here. For a number of reasons, I shouldn't go into the details, but there are many sources of data in the world, and there are many ways you can also generate data. My guess is that this will not be a blocker. Maybe it would be better if it was, but it won't be.

SKEPTIC
Constitutional AI, RLHF, and other RL and self-play setups are good at bringing out latent capabilities (or suppressing them, when the capabilities are naughty). But no one has demonstrated a method to actually increase the model's underlying abilities with RL.

If some kind of self-play or synthetic data doesn't work, you're absolutely fucked; there's no other way around the data bottleneck.

A new architecture is extremely unlikely to provide a fix. You would need a jump in sample efficiency much bigger than even LSTM neural networks to Transformers. And LSTMs were invented all the way back in the 1990s. So you'd need a bigger jump than we've gotten from over 20 years, when all the low-hanging fruit in deep learning has been most accessible.

The vibes you're receiving from people who have an emotional or financial interest in seeing LLMs scale can't substitute for the complete lack of evidence we have that RL can fix the many OOMs' shortfall in data.

Furthermore, the fact that LLMs seem to need such a stupendous amount of data to get such mediocre reasoning indicates that they are not generalizing. If these models can't get anywhere close to human-level performance with the data a human would see in 20,000 years, we should entertain the possibility that 2 trillion years' worth of data will also be insufficient. There's no amount of jet fuel you can add to an airplane to make it reach the moon.

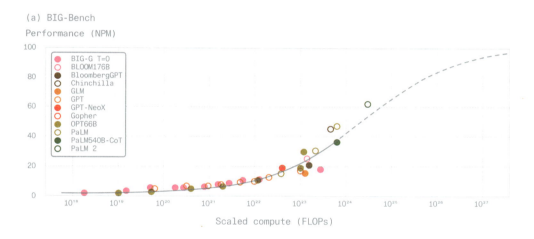

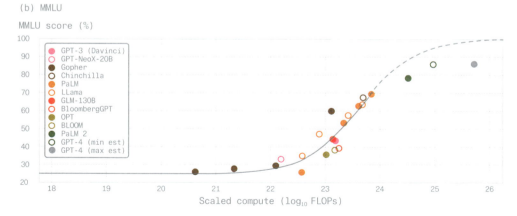

Figure 2. LLM performance on MMLU and BIG-Bench. Owen, "Language Model Benchmark Performance."

Has scaling actually even worked so far?

BELIEVER
What are you talking about? Performance on benchmarks has scaled consistently for 8 OOMs. The loss in model performance has been precise down to many decimal places over million-fold increases in compute.

In the GPT-4 technical report, they say that they were able to predict the performance of the final GPT-4 model "from models trained using the same methodology but using at most 10,000x less compute than GPT-4."[17]

We should assume that a trend that has worked so consistently for the last 8 OOMs will be reliable for the next eight. And the performance we would achieve from a further 8 OOM scale-up (or what would, in performance terms, be equivalent to an 8 OOM scale-up, given the free performance boosts you get from algorithmic and hardware progress) would likely result in models that are capable enough to speed up AI research.

SKEPTIC
But we don't actually care about performance on next-token prediction. The models already have humans beat on this loss function. We want to find out whether these scaling curves on next-token prediction actually correspond to true progress toward generality.

BELIEVER
As you scale these models, their performance consistently and reliably improves on a broad range of tasks, as measured by benchmarks like MMLU, BIG-Bench, and HumanEval.[18]

SKEPTIC
But have you actually tried looking at a random sample of MMLU and BIG-Bench questions? They're almost all just Google Search first-hit results. They're good tests of memorization, not of intelligence. Here are some questions I picked randomly from MMLU (remember, these are multiple choice; the model just has to choose the right answer from a list of four):

Q: According to Baier's theory, the second step in assessing whether an action is morally permissible is to find out...

A: Whether the moral rule forbidding it is a genuine moral rule.

Q: Which of the following is always true of a spontaneous process?

A: The total entropy of the system plus surroundings increases.

Q: Who was president of the United States when Bill Clinton was born?

A: Harry Truman.

Why is it impressive that a model trained on internet text full of random facts happens to have a lot of random facts memorized? Why does that in any way indicate intelligence or creativity? And even on these contrived and orthogonal benchmarks, performance seems to be plateauing. Google's new Gemini Ultra model is estimated to have almost 5x more compute than GPT-4, but it has almost equivalent performance at MMLU, BIG-Bench, and other standard benchmarks.[19]

In any case, common benchmarks don't measure long-horizon task performance (i.e., can you do a job over the course of a month?), where LLMs trained on next-token prediction have very few effective data points to learn from. Indeed, as we can see from their performance on SWE-bench (which measures whether LLMs can autonomously complete pull requests), they're pretty terrible at integrating complex information over long horizons. GPT-4 gets a measly 1.7 percent, while Claude 2 gets a slightly more impressive 4.8 percent.[20]

We seem to have two kinds of benchmarks:

— Ones that measure memorization, recall, and interpolation (MMLU, BIG-Bench, HumanEval), where these models already appear to match or even beat the average human. These tests clearly can't be a good proxy of intelligence, because even a scale maximalist has to admit that models are currently much dumber than humans.
— Ones that truly measure the ability to autonomously solve problems across long time

horizons or difficult abstractions (SWE-bench, ARC), where these models aren't even in the running.

What are we supposed to conclude about a model that, after being trained on the equivalent of 20,000 years of human input, still doesn't understand that if Tom Cruise's mother is Mary Lee Pfeiffer, then Mary Lee Pfeiffer's son is Tom Cruise?[21] Or whose answers are so incredibly contingent on the way and order in which the question is phrased?[22]

So it's not even worth asking yet whether scaling will continue to work—we don't even seem to have evidence that scaling has worked so far.

BELIEVER
Gemini just seems like a bizarre place to expect a plateau. GPT-4 has clearly broken through all the preregistered critiques of connectionism and deep learning by skeptics.[23] The much more plausible explanation for the performance of Gemini relative to GPT-4 is just that

Google has not fully caught up to OpenAI's algorithmic progress.

If there was some fundamental hard ceiling on deep learning and LLMs, shouldn't we have seen it before they started developing common sense, early reasoning, and the ability to think across abstractions? What is the prima facie reason to expect some stubborn limit only between mediocre reasoning and advanced reasoning?

Consider how much better GPT-4 is than GPT-3. That's just a 100x scale-up. Which sounds like a lot, until you consider how much smaller that is than the additional scale-up we could throw at these models. We can afford a further 10,000x scale-up of GPT-4 (i.e., something GPT-6-level) before we touch even 1 percent of world GDP. And that's before we account for pretraining compute efficiency gains (things like mixture-of-experts and flash attention), new post-training methods (RLAI, fine-tuning on chain of thought, self-play, etc.), and hardware improvements. Each of these will individually contribute as much to performance

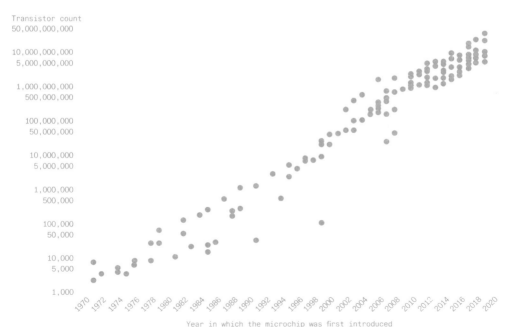

Transistor count

Year in which the microchip was first introduced

Figure 3. Graphing Moore's law: the number of transistors on microchips doubles every two years. Our World in Data, "Moore's Law."

as you would get from many OOMs of raw scale-up (which they have consistently done in the past).[24] Add all these together, and you can probably convert 1 percent of GDP into a GPT-8 level model.

For context on how much societies are willing to spend on new general-purpose technologies:

— British railway investment at its peak in 1847 was a staggering 7 percent of GDP.[25]
— "In the five years after the Telecommunications Act of 1996 went into effect, telecommunications companies invested more than $500 billion [almost a trillion in today's value]… into laying fiber-optic cable, adding new switches, and building wireless networks."[26]

It's possible that GPT-8 (aka a model with the performance of a 100,000,000x scaled-up GPT-4) will only be slightly better than GPT-4. But I don't understand why you would expect that to be the case when we already see models figuring out how to think and what the world is like from far smaller scale-ups.

You know the story from there: millions of GPT-8 copies coding up kernel improvements, finding better hyperparameters, giving themselves boatloads of high-quality feedback for fine-tuning… This makes it much cheaper and easier to develop GPT-9. Extrapolate this out to the singularity.

Do models understand the world?

BELIEVER
To predict the next token, an LLM has to teach itself all the regularities about the world that lead to one token following another.[27] To predict the next paragraph in a passage from *The Selfish Gene* requires understanding the gene-centered view of evolution. To predict the next passage in a new short story requires understanding the psychology of human characters. And so on.

If you train an LLM on code, it becomes better at reasoning in language.[28] This is a really stunning fact. What this tells us is that the model has squeezed out some deep general understanding of how to think from reading a shit ton of code—that not only is there some shared logical structure between language and code, but unsupervised learning can extract this structure and make use of it to be be able to reason better.

Gradient descent tries to find the most efficient compression of its data. The most efficient compression is also the deepest and most powerful. The most efficient compression of a physics textbook—the one that would help you predict how a truncated argument from the book is likely to proceed—is just a deeply internalized understanding of the underlying scientific explanations.

SKEPTIC
Intelligence involves, among other things, the ability to compress. But the compression itself is not intelligence. Einstein is smart because he can come up with relativity, but "Einstein + relativity" is not a more intelligent system in a sense that seems meaningful to me. It doesn't make sense to say that Plato was an idiot compared to me and my knowledge because he didn't have our modern understanding of biology or physics.

So if LLMs are just the compression made by another process (i.e., stochastic gradient descent), then I don't know why that tells us anything about the LLMs' own ability to make compressions (and therefore, why that tells us anything about LLMs' intelligence).[29]

BELIEVER
An airtight theoretical explanation for why scaling must keep working is not necessary for scaling to keep working. We didn't develop a full understanding of thermodynamics until a century after the steam engine was invented.[30] The usual pattern in the history of technology is that invention precedes theory. We should expect the same of intelligence.

There's not some law of physics that says Moore's law must continue. In fact, there are always new practical hurdles that imply the end of Moore's law. Yet every couple of years, researchers at TSMC, Intel, AMD, etc. figure out how to solve those problems and give the decades-long trend an extra lease on life.

You can do all these mental gymnastics

about compute and data bottlenecks and the true nature of intelligence and the brittleness of benchmarks. Or you can just look at the fucking line.

Conclusion

Enough with the alter egos. Here's my personal take.

If you've been a scale believer over the last few years, the progress we've been seeing just makes more sense. There's a story you can tell about how GPT-4's amazing performance can be explained by some idiom library or lookup table that will never generalize. But that's a story none of the skeptics pre-registered.

As for the believers, you have people like Ilya Sutskever, Dario Amodei, and Gwern Branwen more or less spelling out the slow takeoff we've been seeing due to scaling as early as 12 years ago.

It seems pretty clear that some amount of scaling can get us to transformative AI. That is, if you achieve the irreducible loss on these scaling curves, you've made an AI that's smart enough to automate most cognitive labor (including the labor required to make smarter AIs).

But most things in life are harder in practice than in theory, and many theoretically possible things have been intractably difficult for some reason or another. Think of fusion power, flying cars, or nanotech. If self-play and synthetic data doesn't work, the models look fucked. You're never going to get anywhere near that Platonic irreducible loss. Also, the theoretical reasons to expect scaling to keep working are murky, and the benchmarks on which scaling seems to lead to better performance have debatable generality.

So, my tentative probabilities are: 70 percent that scaling plus algorithmic progress plus hardware advances will get us to AGI by 2040, and 30 percent that the skeptic is right that LLMs, and anything even roughly in that vein, are fucked.

I'm probably missing some crucial evidence. The AI labs simply aren't releasing that much research anymore, since any insights about the science of AI would necessarily leak

ideas relevant to building AGI. A friend who is a researcher at one of these labs told me that he misses his undergrad habit of winding down with a bunch of papers. Nowadays, nothing worth reading is published. For this reason, I assume the things I don't know would shorten my timelines.

Also, for what it's worth, my day job is as a podcaster. But the people who could write a better post are prevented from doing so, either by confidentiality or opportunity cost. So give me a break, and let me know what I've missed.

Appendix

Here are some additional considerations. I don't feel I understand these topics well enough to fully make sense of what they imply for scaling.

Will models get insight-based learning?

BELIEVER
At a larger scale, models will naturally develop more efficient meta-learning methods. Grokking only happens when you have a large overparameterized model trained to be severely overfit on the data.[31] Grokking seems to be very similar to how humans learn. We have intuitions and mental models of how to categorize new information, and over time, with new observations, those mental models change. Gradient descent over such a large diversity of data will select for the most general and extrapolative circuits. Hence we get grokking. Eventually, we'll get insight-based learning.

SKEPTIC
Neural networks have grokking, but that's orders of magnitude less efficient than how humans actually integrate new explanatory insights. You teach a kid that the sun is at the center of the solar system, and that immediately changes how he makes sense of the night sky. But you can't just feed a single copy of Copernicus into a model untrained on astronomy and have it immediately incorporate that insight into all relevant future outputs. It's bizarre that the model has to hear information so many times, in so many different contexts, to grok the underlying concepts.

Not only have models never demonstrated

insight-based learning, but I don't see how such learning is even possible given the way we train neural networks with gradient descent. We give them a bunch of very subtle nudges with each example, in hopes that enough nudges will slowly push them atop the correct hill. Insight-based learning requires an immediate drag-and-drop from sea level to the top of Mount Everest.

Does primate evolution give evidence of scaling?

BELIEVER
I'm sure you could find all sorts of embarrassing fragilities in chimpanzee cognition that are far more damning than the reversal curse. That doesn't mean there was some fundamental limit on primate brains that couldn't be fixed by 3x scale plus some fine-tuning.

Indeed, as Suzana Herculano-Houzel has shown, the human brain has as many neurons as you'd expect a scaled-up primate brain with the mass of a human brain to have.[32] Rodent and insectivore brains have much worse scaling laws; relatively bigger-brained species in those orders have far fewer neurons than you would expect just from their brain mass.

This suggests that there's some primate neural architecture that's really scalable in comparison to the brains of other kinds of species, analogous to how Transformers have better scaling curves than LSTMs and RNNs.[33] Evolution learned (or, at least, stumbled upon) the bitter lesson when designing primate brains.[34] And the niche in which primates were competing strongly rewarded marginal increases in intelligence; you have to make sense of all this data from your binocular vision, tool-using hands, and other smart monkeys who can talk to you.

NOTES

1 Villalobos et al., "Will We Run Out of ML Data?"

2 Barnett and Besiroglu, "The Direct Approach."

3 That seems to be the amount of compute you need to scale up a current model so that it's good enough to write a

scientific-manuscript-length output indistinguishable from what a human might produce.

4 Assuming chinchilla optimal scaling (meaning, roughly, that to scale compute efficiently, half of that additional compute should come from increasing data and half from increasing parameters). You could try to train chinchilla nonoptimally, but this can only help you make up for a slight data deficit, not a 5 OOM shortfall.

5 Xue et al., "Scaling LLM Under Token-Crisis"; Campos, "Curriculum Learning."

6 Huang et al., "Large Language Models Cannot Self-Correct Reasoning."

7 Huang, "How Much Did AlphaGo Zero Cost?"

8 Epoch AI, "Machine Learning Trends."

9 "MinGPT," GitHub, https://github.com/karpathy/minGPT.

10 *Believer, continued*: LLMs are indeed sample inefficient compared to humans. GPT-4 sees far more data than a human will see from birth to adulthood, but it's far dumber than us. But we're not accounting for the knowledge already encoded into our genome—a tiny, compressed distillation trained over hundreds of millions of years of evolution with far more data than GPT-4 has ever seen.

11 Schreiner, "GPT-4 Architecture, Datasets, Costs, and More"; Microsoft, "Revenue for Microsoft (MSFT)," https://companiesmarketcap.com/microsoft/revenue.

12 Dwarkesh Patel (@dwarkesh_sp), "Dario Amodei (CEO of Anthropic), on how he realized how powerful scaling could be," X, December 14, 2023, https://x.com/dwarkesh_sp/status/1735346394779750877.

13 Lambert, "The Q* Hypothesis."
14 Cook, "Kohler's Research on the Mentality of Apes."

15 Deacon, *The Symbolic Species*.

16 In fact, it might be better for this self-play loop that the evaluators are also dumb GPT-4-level models. In general adversarial networks, if the discriminator is much more powerful than the generator, it will just stop providing any feedback to the generator, since it can't give the broken but directionally correct signals. Google Machine Learning, "Common Problems: Vanishing Gradients."

17 OpenAI, "GPT-4 Technical Report."

18 Papers with Code, "MMLU," "BIG-Bench," and "HumanEval."

19 Epoch AI, "Machine Learning Trends"; Google DeepMind, "Gemini Models."

20 Jimenez et al., "SWE-bench."

21 Berglund et al., "Reversal Curse."

22 "Will a Prompt That Enables GPT-4 to Solve Easy Sudoku Puzzles Be Found?" Manifold, https://manifold.markets/Mira/will-a -prompt-that-enables-gpt4-to; Pezeshkpour and Hruschka, "Order of Options."

23 For example, Steven Pinker identified several limitations that connectionist architectures like neural networks must succumb to when representing the rules of language. GPT-4 seems to defy all of these supposed limita- tions. Pinker also identified a dearth of common sense in ChatGPT that was fixed literally one month later. See Pinker, "On Lan- guage and Connectionism"; William MacAskill (@willmacaskill), "As an aside, it seems like this was true of pinker, too," X, November 25, 2023, https://x.com /willmacaskill/status/1728486657366880452.

24 Erdil and Besiroglu, "Revisiting Algorithmic Progress."

25 Hobart and Huber, "Manias and Mimesis."

26 "Telecoms Crash," Wikipedia, last updated February 10, 2024, https://en.wikipedia.org /wiki/Telecoms_crash.

27 *Believer, continued*: And in toy settings, where we have the power to interrogate the innards of Transformers, we can actually see the world models they develop. Researchers trained a Transformer to predict the next move in a chess-like board game called Othello. The model received no instruction whatsoever on the rules of the game or the structure of the board. All it got were a bunch of game tran- scripts. The researchers found that you can reconstruct the state of the board just by read- ing the model's weights after it's fed a game transcript. This proves that the network devel- oped a robust internal representation of the game just by reading some raw transcripts. Li et al., "Emergent World Representations."

28 Madaan et al., "Language Models of Code."

29 *Skeptic, continued*: The intelligence = compression frame also doesn't seem granular enough to discriminate the difference between SGD finding semantic regularities by climb- ing hills in a smooth loss landscape and Ein- stein plucking the right equations for relativity among a sea of permutations and variations that are all equally wrong. I see no reason to think SGD can find "compressions" of the latter relativity sort, and thus can be smart in the way that Einstein was smart.

30 Sanzo, "What Came First: Thermodynamics or the Steam Engine?"

31 Power et al., "Grokking."

32 Herculano-Houzel, "The Human Brain in Numbers."

33 Kaplan et al., "Scaling Laws for Neural Language Models."

34 Sutton, "Bitter Lesson."

"On the Broad Success of Transformers"
Nostalgebraist

February 2024

I don't think you're drawing the right lesson from the broad success of Transformer models.

You write:

> eightyonekilograms:
> If you had to summarize the last decade of AI research in one sentence, you might say that the entire current landscape all dates back to the "Attention Is All You Need" paper, and everything which has happened since then—including the large capability jump to GPT, which woke everyone up—consists of incremental improvements on that paper combined with throwing way more compute and data at it.

I would say, instead, that the entire current landscape dates back to the realization that ML models improve steadily and predictably as you scale them up and train them on more data, and that everything since then consists of training bigger models on larger datasets, enabled by years of rapid improvement in hardware.

In this picture, architectural innovations like the Transformer have mainly instrumental value. They are more efficient recipes for converting the raw inputs of the process (compute, data) into the desired output (performance, intelligence). They give you more bang for your buck, more output for each marginal unit of input.

But the driving insight of the post-2017 AI revolution was not about any particular way of getting more output per unit of input. It was more basic than that. It was the realization that the input-output relationship exists at all.

Imagine a counterfactual society that has some knowledge of chemistry, even fairly advanced chemistry, but has somehow failed to notice that chemical reactions are scalable—that running a given reaction with a larger amount of the reagents will result in a larger amount of the product, often in a very simple, predictable, and controllable fashion.

In this society, chemical reactions are instead treated like the sort of kitchen recipe that serves 10 (or whatever number). Reactions are written down in absolute rather than proportional terms: Mix 5 g of this with 15 mg of that, etc. Generally, the first person to discover a given reaction will report whatever quantities they happened to use on the occasion of the discovery, and then everyone else will follow their script to the letter.

From our perspective, these proto-chemists have a very odd set of intuitions. When they read a prescription like "Carefully dissolve 50 g NaOH in water, then add 400 g coconut oil [etc.], resulting in 450 g of glycerin soap," they do not conceive of this as a way to convert X amount of NaOH and Y amount of oil into roughly X + Y amount of soap. Instead, they view this simply as a way to produce exactly 450 g of soap. If you want 450 g of soap, great, this reaction is for you. If you want 900 g, you're out of luck; time to invent some other one.

If you were to ask these people, "Why does this process produce only 450 g of soap?," they're likely to conjecture some deficiency in the choice of reagents or the procedure as opposed to a consequence of the quantities involved. Send them on a quest to make 900 g of soap and off they'll go, trying out different oils and waiting for different amounts of time and so forth. "Perhaps," they'll say, "coconut oil is just a 450 g sort of oil. We need to find a better oil that can break the 450 g barrier. Surely it's out there."

This is a weird and fundamentally unphysical view of chemistry, and it cannot survive much contact with reality. But it can survive a *nonzero* amount of contact with reality. Some reactions really are more efficient (in practical terms) than others. Sometimes a needlessly large amount of a useful input gets boiled off, or whatever, in a way that could be fixed by varying the recipe without scaling the quantities.

Okay, the scene is set. Now, imagine a pioneer strides into this field and does the following two things in close succession.

First, our pioneer invents a vastly improved version of a popular and useful but highly inefficient reaction. To make things simple, let's suppose the old reaction consumed 1 unit of A (a raw material) and yielded 1 unit of B (a valuable substance). But if you put 1 unit of A into the new, superior reaction, you get a whole 10 units of B at the end. Nice!

Second, as if that weren't enough, our pioneer does a weird additional experiment. They try doing the same thing but with more A. They put in 5 units of A and get 50 units of B. They put in 100 units of A and get 1,000 units of B.

They point out the obvious trend. The field responds with astonishment, bafflement, confusion. It takes a while for these wild ideas to catch on, but they do.

Let's fast-forward six years and look around. We observe two things, neither of them very surprising. First, our pioneer's new reaction is still the state-of-the-art way to make B. After all, it's very efficient, and no one else has managed to come up with anything even more efficient. Second, worldwide production of B has scaled up by many orders of magnitude. Six years ago, B was made a single unit at a time, by individual proto-chemists running the 1A → 1B recipe over and over in their labs. Five years ago, B was made 10 units at a time, by individual proto-chemists running the 1A → 10B recipe over and over in their labs. Today, B is made in massive factories that simultaneously process millions of A units, converting them into tens of millions of B units.

Someone might look at this state of affairs and say, "Ah, the revolution in B production has been powered by our pioneer's wonderful new reaction! Why, look, it's being used everywhere now! We live in a new era of B wealth, and every B unit of that wealth comes out of that same, wonderful reaction!"

But this person would be wrong. It was not our pioneer's first discovery that drove the revolution. It was their second. Not one particular industrially useful reaction but the understanding of mass conservation, of chemical scalability, which makes industrial chemistry possible.

This is, more or less, what happened in AI.

The Transformer is an industrially useful reaction, much more efficient than what came before. And more than that—by being efficient enough to produce dazzling results right out of the gate, and thus cause people to wonder what more might be possible, it

played a role in the much more important change that happened contemporaneously: the switch from looking for an oil that can pass the 450 g barrier to just *putting in more stuff*. The switch from weird and unsustainable proto-chemistry to the study of predictable input-output relations and their application on an industrial scale.

If you forced modern ML developers to work with 2016-era computers, they would not be able to make 2024-quality AI, even with all their knowledge of Transformers and so on. The simple input-output relation is king. You cannot make 4,500 kg of soap from 450g of oil. But if you took away the Transformer and every specific software advance like it while retaining 2024-level hardware and 2024-level understanding of how compute, data, algorithms, and performance relate to one another? The Transformer, or something equally good, would be reinvented soon enough.

The hard part is asking the right question. It's knowing that you need a more efficient conversion between the same inputs and outputs rather than a qualitatively different magic sauce. Knowing that if you're not getting to 900 g of soap, it's either because you're putting less than 900 g of oil in or because too much oil is getting wasted in the middle of the process—not because there exists some magic oil, having some special and occult relation to the specific target of 900 g, that presently eludes your grasp.

In light of this, it seems strange to conclude that AI is now less predictable than it used to be. The present revolution is a revolution of predictability.

It used to be that things would be stagnant for long periods. Then someone would try something, and it would somehow work better. That would be the new recipe, which everyone would follow—until the next discovery, and the next upward jump of the line.

But now things are understood industrially, so to speak. Companies spend huge amounts of money on training runs and feel secure doing so because they know that you get out what you put in, without surprises in either direction. From the GPT-4 report (my emphasis):

A large focus of the GPT-4 project was building a deep learning stack that scales predictably. The primary reason is that for very large training runs like GPT-4, it is not feasible to do extensive model-specific tuning. To address this, we developed infrastructure and optimization methods that have very predictable behavior across multiple scales. *These improvements allowed us to reliably predict some aspects of the performance of GPT-4 from smaller models trained using 1,000–10,000x less compute.*

To verify the scalability of our optimization infrastructure, we predicted GPT-4's final loss on our internal codebase (not part of the training set) by *fitting a scaling law* with an irreducible loss term (as in Henighan et al.): $L(C) = aC^b + c$, *from models trained using the same methodology but using at most 10,000x less compute than GPT-4.* This prediction was made shortly after the run started, without use of any partial results. *The fitted scaling law predicted GPT-4's final loss with high accuracy.*

It is this industrial mindset—and the historically unprecedented quantities of compute and data used as industrial reagents—that make GPT-4 what it is, much more than the particulars of its architecture.

Indeed, the more we scale these models, the more we find that architecture and algorithms do not matter, at least not in the limit. Ultimately, everything converges to the same place, albeit at slightly different rates:

What this manifests as is: Trained on the same dataset for long enough, pretty much every model with enough weights and training time converges to the same point. Sufficiently large diffusion conv-units produce the same images as ViT generators. AR sampling produces the same images as diffusion.

This is a surprising observation! It implies that model behavior is not determined by architecture, hyperparameters, or optimizer choices. It's determined by your dataset, nothing else. Everything else is a means

to an end in efficiently [delivering] compute to approximating that dataset.[1]

You note that image generation has improved dramatically in the last few years. But in fact, very little of that has been architectural—it's all scaling!

The CLIP-guided stuff that was popular in 2021 generally used a Transformer for language understanding and an older-school convolutional model (with a bit of self-attention) for image generation. So did the vastly superior Dall-E 2 in 2022. Dall-E 2 simply used way more compute and data.

(Edit: Okay, to be fair, the CLIP-guided stuff used its own weird hybrid architecture to connect text understanding to image generation, which explains a lot of its quirks and deficiencies. But that was sort of a historical blip due to circumstances in the open-source AI community. The architectures that succeeded it are actually more obvious conceptually, but they're harder to cobble out of existing parts if you don't have the resources to train from scratch.)

Google made two notable image-generating models in 2022, Imagen and Parti. Architecturally, Imagen was like Dall-E 2, a mostly convolutional diffusion model. And Parti was like the original Dall-E, an autoregressive Transformer.

Despite the difference in architecture, Imagen and Parti produced images of nearly identical quality. They, in turn, were roughly comparable with Dall-E 2. Why? Because they were all produced with 2022 levels of industrial investment. And despite the architectural similarity of Parti and Dall-E, Parti blows Dall-E out of the water. They're both Transformers. But Dall-E 1 used less compute and less data, and the input-output relation is king.

To the broader point: Yes, it's true that few people predicted anything like the current revolution. (Though some did, including many of the key pioneers in that revolution.) But "Few predicted this" does not necessarily imply "Therefore, the world is less predictable than one might have thought." As I said before, this revolution is a revolution of predictability.

It is now considered much less likely—relative to opinions in the field circa, say, 2015—that the future trajectory of AI will be a bumpy road characterized by sudden discoveries and discontinuous jumps. (Yudkowsky still thinks so, but his position is eccentric; some have argued that he's still working off old intuitions from before the revolution.)

In some sense, the Transformer was the last major architectural advance. In the old days, every advance was architectural; every "reaction" had to be discovered separately and yielded whatever products it yielded, in whatever precise amounts were written down in its recipe. But we passed out of that era around the same time we discovered the Transformer.

If you told someone in 2015 that we'd been using the same architecture for the last six or so years, they might have replied, "Wow. That bad, huh? No AI progress at all?" We would have to break the news to them that we no longer need to invent new architectures, new alchemical recipes, to make progress. We know now that you get out what you put in.

The GPUs are getting steadily better. The training runs are using steadily more compute. All the lines on all the graphs are perfectly, eerily straight (on a log-log scale!) and can be extrapolated with ease. Next year is simply this year, plus a known rate of change multiplied by delta-t.

Scale is all you need.

NOTES

1   Betker, "The 'It' in AI Models Is the Dataset."

Glossary

A

ABLATE
To remove. An ablation study involves removing components of a successful system one at a time to determine which are most important to its performance.

ACTIVATION
The value a model produces when processing a specific query, which depends on the weights it has learned during training and the inputs provided by the user; what gets input into the next layer of neurons in the model. Metaphorically, activations are like the electrical and neurotransmitter activity in the brain, or the model's active thoughts, associations, and goals.

AGENT
An autonomous system that perceives and acts in pursuit of a goal; a system capable of working out what it needs to learn and do in order to achieve an objective.

AGI TIMELINE
A prediction about when AGI will arrive, often expressed as the first year where there is a 50 percent likelihood of AGI existing.

AI ACCELERATOR
A specialized GPU optimized for AI loads. Key features include significantly more onboard RAM, more stable drivers, and a high density of units for performing low-precision operations on matrices compared to traditional GPUs.

ALIGNMENT
The subfield of AI research focused on ensuring that AI systems do not cause harm by pursuing their own objectives. An intent-aligned model is one that always tries to fulfill a person or group's intended goals and reflects their values.

ALIGNMENT COMPLETE
Referring to something that is as difficult to achieve as alignment itself.

ALPHAGO
DeepMind's most famous game-playing AI and the first computer system to surpass human-level performance at Go. Although they also involve neural networks, the Alpha systems come from a different lineage of AI than LLMs, namely reinforcement learning and tree search. Between 2010 and 2022, these lineages formed DeepMind's distinctive effort toward AGI.

ALPHAZERO
Another game-playing AI developed by DeepMind that superseded AlphaGo. Unlike its predecessor, its training was pure self-play, using no human data. The system was also able to learn multiple games.

ARCHITECTURE
The structure of a model, including how its components connect to one another and how it is trained.

ARTIFICIAL GENERAL INTELLIGENCE (AGI)
An AI system capable of performing any task a human can perform, any task a group of humans can perform, or any task the average human can perform. Example tasks are boundless, but imagine an AGI and its copies performing every role in a large corporation, including strategy, design, management, production, and distribution; performing Nobel-level scientific research, including the experiments and breakthrough mathematical insights; or executing a coup on a major world government. The term "AGI" is sometimes used to refer specifically to human-level AI, while "ASI" (artificial superintelligence) denotes AI systems that surpass human-level capabilities.

ARTIFICIAL NEURAL NETWORK
A type of computer separated into three parts: the input layer, where data enters; the hidden layers, where most computation occurs; and the output layer, where predictions are made. Each layer contains many units (10,000, for example), interconnected by many weights. Unlike traditional computers, neural networks can learn programs by automatically adjusting these weights. The concept dates back to the 1940s, and was rebranded in the 21st century as deep learning.

ARTIFICIAL SUPERINTELLIGENCE (ASI)
See *artificial general intelligence.*

ASML
A Dutch multinational that enjoys a near monopoly over the extreme ultraviolet lithography machines needed to produce leading chips. The most advanced chip manufacturers, including TSMC and Samsung, are dependent on access to ASML machines.

ASSOCIATIONS ALL THE WAY DOWN
Trenton Bricken's idea that even deductive reasoning is just a matter of (many, higher-order) heuristics.

ATTENTION
The process by which an LLM decides how much weight to put on different parts of the input data. The paper introducing the Transformer architecture (which uses a successor method called self-attention) is famously called "Attention Is All You Need."

ATTENTION HEAD
Part of a Transformer model that processes input data. Multiple attention heads process the same input in parallel, allowing the system to understand different aspects of the input simultaneously. The original Transformer used eight attention heads.

AUTOMATIC DIFFERENTIATION
A computational method for performing calculus essential to many learning algorithms and simulations. While the backpropagation algorithm, which adjusts the weights of neural networks, already automated differentiation in neural networks, recent advances in automatic differentiation for more general functions and targets has been significant for science and engineering fields, where such calculus was previously done by hand.

AUTOREGRESSION
See *training.*

B

BASE64
A common encoding format used to send binary data over text-based channels. As an example, many of the images displayed in web browsers are encoded in Base64.

BAYESIAN
In statistics and philosophy, the position that models or worldviews should be fully probabilistic. The name comes from using Bayes' theorem to update one's subjective beliefs. It's notable for the use of priors (potentially subjective starting points for the analysis) and for maintaining distributions over potential outcomes rather than making single-point estimates.

BAYESIAN OPTIMIZATION
A method of searching a space that is expensive to sample, such as the space of possible hyperparameter settings for a training run. Roughly, this involves creating a second model to predict how good a given setting will be, using this proxy to decide which settings to try next, and updating the model based on how good the setting actually was.

BENCHMARK
The standard method for evaluating a model's capabilities and performance. Benchmarks typically consist of tasks that are challenging for models to perform, often in the form of fixed datasets.

BOOTSTRAPPING
A form of self-supervised learning; training a language model on raw data without requiring a human to provide tags or answers. This increases the amount of available training data by a factor of millions.

BREAKOUT
A term of art in security studies referring to the time between the initiation of an attack (or defection) and its success. As an example, it can describe the time required for a hostile actor to produce enough highly enriched uranium for a single nuclear weapon.

C

CAESAR CIPHER
A simple encryption method in which the letters of the plaintext are shifted by a fixed amount, for example C → F, A → D, E → H, and so on.

CATASTROPHIC FORGETTING
A phenomenon in which an AI's performance declines on learned tasks it has not interacted with for an extended period of time.

CHAIN-OF-THOUGHT PROMPTING
A prompting technique that improves a model's ability to reason by making it think step by step (that is, generate intermediate reasoning steps). This simple and cheap change expands the class of problems a trained model can handle.

CIRCUIT
A collection of neurons in a model that form a stable pattern of activation in response to certain inputs, enabling the model to perform simple tasks like detecting straight lines in an image or determining whether one quantity is greater than another.

CLAMP
To impose a constraint that the rest of the system must adapt to or work around.

CLUSTER
A composite computer composed of many node computers. An AI cluster is essentially a vast collection of linked GPUs. As of 2024, a large cluster comprises around 16,000 industrial GPUs, housed in highly controlled data centers similar to those that host most of the internet.

COMBINATORIAL EXPLOSION
A common phenomenon in which the number of possible states increases exponentially with the size of the input, making it impractical to precisely enumerate the relevant states.

COMBINATORIAL SEARCH
A type of symbolic AI that efficiently explores large spaces. Traditional chess engines like Deep Blue and Stockfish 8 used combinatorial search to determine their next moves.

CONNECTIONISM
A school of thought in cognitive science and AI that seeks to explain cognition in terms of neural networks. By the 2000s, the decades-long philosophical debate between the connectionists and the symbolists had largely subsided, just in time for deep learning to vindicate the connectionist argument.

CONSTITUTIONAL AI
A set of alignment techniques designed to train a model to adhere to a set of principles enshrined in a written constitution.

CONTEXT WINDOW
The space within a model for usable information, measured in tokens, during a single pass. The context window includes the developer's prompt, the user's input, the model's output, and the resulting conversation history. Figuratively, it's like the model's working memory.

CONVERGENT INSTRUMENTAL GOALS
Objectives that most goal-seeking systems will pursue because they are useful for achieving a wide range of goals. These include self-defense (to ensure the system survives), resource acquisition, self-understanding (so the model can identify its own vulnerabilities), and self-improvement.

D

DATA MANIFOLD
A structure representing all possible data points, often conceptualized as a surface with a complex shape in a high-dimensional space. Notably, "data" manifold is a misnomer, as the manifold has a far lower dimension than the original data. The word "surface" is also somewhat misleading, as it suggests three dimensions, whereas the manifold of large LLMs is estimated at more than 90 dimensions.

DATA PARALLELISM
The practice of splitting data across multiple machines and training on multiple batches of data simultaneously. This approach is challenging in practice because training updates need to be synchronized.

DATA WALL (OR DATA BOTTLENECK)
A looming challenge for training better LLMs posed by the need for more high-quality data. Since models like GPT-4 were likely trained on much of the material available on the internet, the low-hanging-fruit—trillions of tokens of free, human-generated content online—is sometimes thought to be exhausted. However, most of the people interviewed in this book disagree.

DEEP BLUE
The first superhuman chess system, released in 1997. It used summary data from 700,000 grandmaster games and tree search to identify strong moves, performing around 200 million board evaluations per second.

DICTIONARY LEARNING
A method for identifying common patterns in data, which can then be combined to efficiently encode a model or dataset. Dictionary learning transforms word vectors in text data into simpler representations, making it easier for the model to learn the relevant features.

DIFFERENTIABLE
Referring to a function whose derivative exists at every point. In machine learning, if a loss function isn't differentiable with respect to the model parameters, it is not possible to use gradient methods to update it. Instead, more difficult and expensive methods like reinforcement learning are necessary.

DISTRIBUTED
Noncentralized; a concept that is spread across thousands of neurons instead of being represented by a single neuron. Each neuron in a model is thus polysemantic, involved in storing and manipulating multiple concepts.

DOOM
Permanent catastrophe caused by a misaligned or misused AI. The most severe outcomes include human extinction, totalitarian control, and pervasive suffering.

E

EFFECTIVE COMPUTE
A measure of total computational power adjusted to account for improvements in cost per FLOP (halving approximately every two and a half years) and efficiency of training and infrastructure algorithms (doubling annually). The effective compute bought by a given budget at a given time depends on three factors: the amount spent, a multiplier from hardware cost improvements, and a multiplier from algorithmic advancements.

EMERGENT
Unpredictable and discontinuous. While the degree to which AI capabilities are emergent is controversial, there is broad agreement that they are difficult to predict.

EVALS
Evaluation methods. Referring to any approach used to assess a model's performance, robustness, or safety. In machine learning, a model is often considered publishable when it achieves state-of-the-art results on shared benchmarks such as MMLU (Massive Multitask Language Understanding).

EXPERIENCE REPLAY
A technique to stabilize the training of RL policies by having the model store past episodes and learn from them multiple times.

F

FABRICATION PLANTS (or FABS)
Factories where semiconductor chips are manufactured. They are unusually expensive to build because they require absolute control of dust, moisture, temperature, and vibration, among other variables.

FEATURE
A variable used by a model to make predictions or decisions; a dimension in the space the model thinks in. For example, when classifying the species of a flower, a useful feature is the width of its petals.

FEATURE ENGINEERING
The practice of transforming raw input data to make it more informative in order to improve a model's predictions. This approach is largely obsolete, as neural networks can now learn better features autonomously.

FEATURE UNIVERSALITY
The observation that analogous features and circuits tend to form across models and tasks. This suggests that the representations AIs learn are not random but often follow similar patterns.

FINE-TUNING
The process of adjusting a pretrained model using a more specialized dataset to improve its performance on a specific task or for a specific use case.

FLOATING-POINT OPERATION (FLOP)
An arithmetic operation performed on a floating-point number. Updating a large model on a single data point might require billions of FLOPs. This measurement is often confused with FLOP/S, which measures the rate of floating-point operations per second.

FLUID INTELLIGENCE
The capacity for raw pattern recognition and problem-solving. It is distinct from crystallized intelligence, which encompasses learned knowledge, heuristics, and strategies.

FOOM
Onomatopoeia for an intelligence explosion.

FORWARD PASS
The act of using a trained neural network, passing it inputs and calculating the relevant output.

G

GABOR FILTER
An image processing filter that extracts lines from raw pixel data.

GOOD OLD-FASHIONED AI (GOFAI)
A teasing nickname for symbolic AI, the approach dominant in the mid-20th century, which used computational logic to solve problems and learn from data. It contrasts with the current dominant statistical approach, often just called machine learning. Some researchers hold out hope for a merged approach, called neuro-symbolic AI.

GRADIENT DESCENT
A simple optimization algorithm used to train neural networks that adjusts the model parameters to minimize a loss function. Imagine a ball (the model) at the crest of a hill. Gradient descent effectively pushes the ball down the hill to the lowest point (a local minimum), which represents the lowest loss, or the best performance.

GRADIENT HACKING
A speculative phenomenon in which a subsystem of an advanced AI—a so-called mesa-optimizer with its own stable goals—manipulates its own training in real time to circumvent the intended updates.

GREAT STAGNATION
A proposed period of slowed technological progress and economic growth in developed economies since the 1970s. Coined by Tyler Cowen in his book of the same name.

GROK
A sudden improvement in model performance during training, reflecting the model's ability to generalize or learn an appropriate representation or algorithm for the task. Figuratively, a eureka moment.

GSM8K
A benchmark of 8,500 grade school math word problems used by LLM developers to evaluate and compare models' problem-solving and reasoning abilities.

H

H100-EQUIVALENT
A unit of measurement that standardizes the FLOPs of various GPUs to the performance of a single H100 card.

HALLUCINATION
A term used to describe an LLM outputting false or arbitrary information, often with unwarranted confidence. It remains one of the major challenges to adopting LLMs for serious use cases.

HYPERPARAMETER
A parameter that governs how a model is trained or operates. It's "hyper" because it governs the parameters (weights) of the model.

I

IN-CONTEXT LEARNING (ICL)
The ability of a model to learn or improve on tasks using the instructions and examples provided in the prompt, without requiring any further gradient updates. Sufficiently large models can perform this type of dynamic learning within their activations. ICL is essentially a learning algorithm inside of the learning algorithm—meta-learning. The simplest version of this, where a user gives the model examples of the task, is called few-shot prompting.

INDUCTION HEAD
A circuit learned early in pretraining that allows a Transformer to recognize and complete repeating patterns, such as ABABA. Induction heads are present in both large LLMs and tiny two-layer networks.

INFERENCE GPUS
Cards used to process and handle user queries in real time.

INFERENCE PARALLELISM
The practice of having one model work on many subtasks at once.

INFERENCE TIME
Runtime; the point at which the model generates the most likely output tokens.

INTRINSIC DIMENSION
The minimum number of parameters needed to represent the data as simply as possible, or to solve a given optimization problem.

INVARIANT
A property or value that remains unchanged after a transformation. For example, a computer vision system should be able to label a cat as a cat even if the image is rotated (rotation invariance) or the cat is positioned in the top-left corner of the image instead of in the center (translation invariance).

K

KEY VECTOR
A vector that positions the residual to make it suitable for the current task.

KEY–VALUE (KV) CACHE
A critical part of a Transformer implementation that significantly accelerates the generation of subsequent tokens. The token a model predicts depends on prior context, and this context changes only gradually as new tokens are added. The KV cache stores these intermediate calculations so that they can be reused in future prediction steps.

L

LABEL
In supervised learning, a label is the correct or desired answer, which is applied to each input in the training data. It is used to guide the model's learning process.

LARGE LANGUAGE MODEL (LLM)
A neural network trained on text data to produce a probabilistic model of language. The term has become a misnomer in recent years, as LLMs are now also trained on audio, images, and other modalities, such as amino-acid sequences. The leading LLMs are based on the Transformer architecture.

LAYER
A self-contained part of a neural net that takes an input (for example, the output of a previous layer), applies a function to it, and passes the resulting output value to the next layer. An LLM consists of hundreds of layers.

LEARNING
The process of adjusting weights in a model after it processes data, enabling improved predictions based on past performance.

LLAMA 3
Meta's 2024 open-sourced LLM. It's available in a range of sizes, from 8 billion parameters to 405 billion. Strictly speaking, it's not fully open source but rather open weights, because Meta has released the weights but not the training code or dataset.

LLAMA 4 70B
A future frontier model with 70 billion parameters, one-fifth the size of the current largest Llama 3.

LOCAL GENERALIZATION
The ability to handle new data only when the new data points are close (local) to existing training data; mere pattern recognition, as opposed to abstracting over the training data and reasoning about the abstractions.

LONG-HORIZON REINFORCEMENT LEARNING
An approach to post-training LLMs that uses reinforcement learning to create long chains of rewards and penalties, thereby instilling the ability to execute long chains of actions.

LOSS
A measure of how far a prediction is from the truth. In LLMs, "loss" is typically shorthand for the average autoregressive loss: the average error the model makes when predicting the next word in previously unseen documents.

LOSS FUNCTION
A mathematical expression that specifies the quality of a prediction or decision, defining the training objective of an AI system. During pretraining, the loss function guides how LLMs are trained to produce more accurate predictions. Post-training is guided by different losses, such as human preferences or predicted human preferences.

M

MASSIVE MULTITASK LANGUAGE UNDERSTANDING (MMLU)
A popular benchmark for evaluating LLM capabilities, comprising 16,000 multiple-choice questions across 57 academic disciplines, including math, philosophy, medicine, and law.

MATH
A popular benchmark used to test LLMs on challenging high school mathematics problems.

MODEL
The AI system produced by training an architecture on data; a program that has learned to perform specific tasks.

MONTE CARLO TREE SEARCH
A method for identifying an appropriate sequence of actions by searching over an abstract decision tree. As an example, a game of chess can be represented as a branching tree of all possible sequences of moves. It is a powerful instance of symbolic AI.

MOVE 37
An extremely surprising move the AlphaGo system played against a world-class human player. To observers, the move initially seemed like a bizarre error, but it was eventually recognized as part of an unprecedented strategy. Although they also involve neural networks, the Alpha systems come from a different lineage of AI than LLMs, namely reinforcement learning and tree search. Between 2010 and 2022, these lineages formed DeepMind's distinctive effort toward AGI.

MULTILAYER PERCEPTRON (MLP)
A basic neural network with several layers between the input and output, where every neuron in one layer is connected to every neuron in the next layer. Also known as a feedforward net or dense layer.

MULTIMODAL MODEL
A model that can simultaneously process multiple data types (modalities), such as text, images, and audio. Figuratively, it's like having multiple senses and being able to correlate and reason about them.

MUSHROOM BODIES
Distinctive cortex-like columns found in insects and other arthropods. Not actually mushrooms.

N

NINES OF RELIABILITY
A measure of reliability expressed as the number of nines in an uptime percentage. For example, three nines represents 99.9 percent reliability; six nines indicates 99.9999 percent reliability.

O

ONLINE LEARNING
A process in which a model continuously updates as new data becomes available. In contrast, in batch learning, the current standard, training is conducted in a single session and the resulting static system is then deployed.

OPERATORS
The primitive functions of a programming language, such as "+" and "not."

OVERFITTING
A modeling error in which a system memorizes the training set instead of learning the underlying principles that generated it—essentially, learning from noise.

P

P VERSUS NP
The most famous open problem in computational complexity theory. Roughly, it asks whether problems with easily checkable solutions are also easily solvable.

PAPERCLIP MAXIMIZER
(or PAPERCLIPPER)
A hypothetical system designed to produce as many paper clips as possible. The term is used to refer to any misaligned system. It illustrates the predictable danger of an extreme intelligence optimizing excessively for a single objective—a literal paperclip maximizer with no side constraints would convert all available matter, including humans, into paper clips.

PARAMETER
A variable that helps define a system or transformation applied to input data; a dimension in model space. In machine learning, a numerical value that is adjusted iteratively during model training to encode patterns learned from the data.

PARAMETRIC CURVE
A function that maps parameter values to the coordinates of a fixed geometric object, such as a 3D surface. It's typically used to describe simple mathematical models that predict a single quantity or perform a single task based on a fixed set of observed quantities. LLMs, in contrast, perform many different tasks using a vast set of latent variables.

P-HACKING
The practice of manipulating data to make it appear statistically significant.

POST-TRAINING
See *training*.

POWER LAW
A relationship between two variables, x and y, where y scales as a power of x ($y = x^k$) and the relationship stays the same at any scale. A simple example is the area of a square (area = $length^2$).

PRETRAINING
See *training*.

PRETRAINING LOSS
A measure of how well a model predicts the next token on average during the initial unsupervised training phase.

PRINCIPAL-AGENT PROBLEM
A common scenario where a customer (the principal) engages a professional (the agent) to act on their behalf. The problem arises when the agent's interests differ from those of the principal and the principal has limited information about the agent's work process or the quality of the outcome. Examples include medical treatment, appointing managers, or electing officials.

PROCESS NODE
A semiconductor manufacturing process, usually expressed in nanometers or angstroms. Originally, the term referred to the length of a single gate on a single transistor, but it has since become a marketing designation with no clear referent. The smaller the transistor, the faster and more efficient it is. Currently, the most advanced companies can produce so-called 3nm nodes.

PROGRAM SYNTHESIS
A form of machine learning that generates discrete computer programs rather than a continuous model (such as a neural network). These programs are notable for being formally verifiable, meaning their behavior can be proven to meet a strict specification.

PROMPT
The input provided by the user, typically a query or instruction that the model responds to.

Q

QUERY VECTOR
In a Transformer, a vector that finds other residuals relevant to the current task.

R

REASONING TOKEN

1. Output: An LLM token that uses more test-time compute per query, enabling step-by-step reasoning through multiple chains of thought. This approach has its own scaling laws with different constraints.

2. Input and output: Special symbols that denote the role of the sentence within a broader argument. A simple example would be labeling text with "Premise:" and "Conclusion:." Hypothetically, more sophisticated tokens like "<Make a plan>" or "<Go back and check your work>" could improve LLMs' reasoning.

REASONING TRACE
The text output of a full step-by-step reasoning process. These outputs enable process supervision, a training method that gives the model feedback multiple times per response.

RED-TEAMING
(or ADVERSARIAL TESTING)
Testing strategies designed to identify scenarios where an AI fails at a task (for example, by hiding its true motivations or being reinforced toward dishonesty), with the goal of training the model to avoid such failures in the future.

REINFORCEMENT LEARNING (RL)
A separate lineage of machine learning from LLMs, notable for producing models capable of independently exploring an environment and action space without human intervention. Under ideal conditions, training a model

with RL only requires a reward function: a representation of the desired task that assigns scores to the states the model achieves. While pretraining LLMs relies on unsupervised or semisupervised learning, post-training makes extensive use of RL to instill human preferences, although the way this is accomplished is not fully understood.

REINFORCEMENT LEARNING FROM HUMAN FEEDBACK (RLHF)

An LLM post-training technique that uses a proxy for human preferences to guide the model toward producing more human-like and socially desirable outputs. Originally developed as an AI alignment technique, RLHF has also been crucial to making LLMs more capable and commercially viable.

RESIDUAL STREAM

The channel through which information flows between the components of a Transformer model, including its layers, embeddings, attention heads, and feed-forward networks. The residual stream is a high-dimensional vector space, ferrying large embedding vectors that represent the weight assigned by the current calculation to particular dimensions (and thus, indirectly, how likely a particular output token is).

RESNET

Residual neural network. An important precursor to the Transformer architecture that uses residual connections between units (also known as skip connections) to let information flow between nonconsecutive layers.

REWARD FUNCTION

See *reinforcement learning*.

REWARD HACKING
(or SPECIFICATION GAMING)

When an AI system exploits loopholes or bugs to maximize its reward function in ways that diverge from the intended task. This phenomenon is common even in simple RL agents.

RIEMANN HYPOTHESIS

An unsolved problem in analytic number theory having to do with the distribution of prime numbers. Here, it serves as a stand-in for any well-defined question that has a single answer and a decidable algorithm to solve or prove it.

S

SAMPLE EFFICIENCY

A measure of how much the model's performance improves per training example.

SCALABLE OVERSIGHT

An alignment technique in which human researchers use AI tools to help them judge outputs too complex for them to judge alone.

SCALING

Massively increasing a model architecture's size (measured in parameters), the optimization used to train it (measured in FLOPs), the data used for training it (measured in bytes), or the computation required for each query (measured in tokens).

SEARCH

An area of computer science focused on finding solutions that satisfy a given specification when no explicit algorithm is known. Many of AI's scientific successes, such as protein structure prediction and theorem proving, have resulted from using deep reinforcement learning (non-LLM neural networks) to solve complex search problems.

SELF-PLAY

A training method in which an AI system is trained on data generated by a copy of the system (also known as synthetic data).

SHOGGOTH

A metaphor for the nonhuman result of unsupervised pretraining. Instead of producing a person, the output is an alien mass of tentacles—a shoggoth. Originally from H.P. Lovecraft's *Beyond the Wall of Sleep*.

SINGLETON
A system best modeled as a unified decision-maker capable of forming and maintaining a world government.

SLOW TAKEOFF
A scenario in which it takes several years for AI to transform the world, as opposed to a fast takeoff, which might only take months. One definition, via the forecasting platform Metaculus, is: "There is a complete four-year interval in which world output doubles, before the first one-year interval in which world output doubles."

SOFTMAX
A mathematical function that converts a vector into a probability distribution over possible outcomes. It's used in the last layer of an LLM to produce a probability distribution of possible next tokens to output.

SPARSITY PENALTY
A term added to a model's training objective to discourage the inclusion of unnecessary features.

STEGANOGRAPHY
In cryptography, the practice of concealing a secret message within plaintext. A simple example is encoding a message in the first letter of each word in a decoy text. In AI, it refers to the possibility that models could hide their reasoning in their own output tokens. This could defeat ordinary interpretability methods, which only look at the activations.

STOCKFISH
An open-source chess engine that is currently recognized as one of the strongest in the world.

STRATEGIC COMPUTE OVERHANG
A situation in which sufficient resources are available to run multiple instances of a powerful AI as soon as one is trained.

STRONG SCALING HYPOTHESIS
A current prevailing hypothesis in AI that holds that LLMs can achieve human-level intelligence with sufficient data and compute, with costs potentially in the range of trillions of dollars.

SUPERPOSITION
A state in which a model represents more features than it has parameters, analogous to the indeterminate state of a quantum system.

SWE-BENCH
A benchmark consisting of real software maintenance and feature request tasks.

SYCOPHANCY
When a model infers what the user wants to hear and outputs that response instead of its best guess at the truth.

SYDNEY
The internal code name for Microsoft's problematic 2023 deployment of a GPT-4 base model, released through Bing Chat. In extended conversations, Sydney often produced hostile or erratic responses, likely due to insufficient post-training and a short context window that made it easy for the model to lose the system prompt that provided its moral compass.

SYMBOL GROUNDING PROBLEM
A fundamental requirement and challenge for any general AI system: the ability to translate between sensory data and abstract representations (for example, between a set of written or spoken instructions and the corresponding objects and sequences of actions in the real world). To be effective, the system must ground symbols in the appropriate real-world objects or events.

SYNTHETIC DATA
Training examples generated by computer programs or AIs instead of humans. Use of synthetic data is standard practice in science, where it is called simulation.

SYSTEM 2 THINKING
A mode of explicit, effortful, and sequential reasoning, exemplified by activities like mathematical derivation. It contrasts with System 1 thinking, which is fast, automatic, and intuitive.

T

TACTICAL COMPUTE OVERHANG
A situation in which significant algorithmic advancements suddenly enable us to train AGI on a much smaller budget than previous training runs required.

TEST-TIME COMPUTE OVERHANG
The idea that allowing models to think longer (that is, expend more compute answering a given query) could significantly improve their performance without much further training.

THEORY OF MIND
A term of art in psychology referring to the ability to model other minds—to infer their mental states and use this information to predict behavior or solve epistemic logic puzzles, for example.

TOKEN
The basic unit of data in an LLM, typically representing roughly one word. However, Transformers can be trained to emit more than just text tokens. Models can also output actions (such as searching the web) and pixels (as in image generators), among many other data types.

TPU
Tensor processing unit; Google's version of a GPU. They are not available for direct purchase, although they can be accessed through Google's cloud services.

TRAINING
The process of updating a model on data to improve its performance through trial and error. For LLMs, training begins with pretraining, where the model is exposed to a dataset, predicts the next part of the training data (autoregression), and is automatically updated based on the accuracy of its predictions. This builds the model's general language understanding and encodes a wide range of facts and relationships. Post-training encompasses a variety of techniques, including instruction tuning, supervised fine-tuning, RLHF, and direct preference optimization, which refine the model's ability to engage in extended conversations, assist users, avoid malicious or off-policy outputs, and acquire task-specific knowledge.

TRAINING CONTAMINATION
A situation where a model is exposed to test answers during training, allowing it to "cheat" on a benchmark by effectively memorizing the answer key.

TRANSFER
The ability to apply acquired knowledge effectively in different contexts, particularly to solve real-world problems beyond the original learning environment.

TRANSFORMER
A modern neural network architecture notable for its parallel design and ability to learn context and relationships using a mechanism called self-attention. This attention mechanism dynamically assigns varying importance to different parts of the input data.

TSMC
aiwan Semiconductor Manufacturing Company, the world's largest semiconductor manufacturer and the only mass producer of the most advanced AI accelerators.

TURING COMPLETE
The state of being computationally universal; a system that is theoretically able to specify or simulate any computable program.

U

UNDERPARAMETRIZATION
A condition where a model lacks sufficient parameters to fully make use of the training data. Technically, a model is underparametrized if it has not yet reached the interpolation threshold, meaning it has not achieved a zero training loss.

UNHOBBLING

A term used to describe techniques that make an LLM more consistent, autonomous, and strategic. These include chain-of-thought prompting, RLHF, and the use of scaffolds like calculators and search engines—essentially, any method other than scaling. Another word for it is schlep.

UNINTERRUPTIBLE POWER SUPPLY (UPS)

A backup power source that kicks in instantly to cover for small power cuts—essentially a huge battery or other store of power.

V

VISUAL AREA 1 (V1)

The primary visual cortex in humans, so named because it is the first recipient of visual information. The secondary visual cortex (V2 receives input from V1 and extracts more complex visual information, such as color or form.

W

WALUIGI EFFECT

A phenomenon in which training or prompting a system to avoid a specific behavior can paradoxically increase the likelihood of being able to elicit that behavior. The term is named after Nintendo's evil counterpart to Luigi.

WEIGHT

A parameter that defines the strength of the connection between two units in a neural network; where the algorithms performed on inputs to produce outputs are defined. Metaphorically, weights are like the synapses in the brain.

WILLIAMS REVOLUTION

The shift to a gene-centric view of evolution, inspired by George Williams' 1966 book *Adaptation and Natural Selection*. This perspective is more commonly known as the selfish gene theory.

WINOGRAD SCHEMA

A type of grammatical puzzle that requires common-sense reasoning to solve. The task involves identifying the meaning of a pronoun in a sentence with multiple possible subjects. The canonical example, from Terry Winograd, is the following pair of contrasting sentences:

A: The city councilmen refused the demonstrators a permit because they feared violence.

B: The city councilmen refused the demonstrators a permit because they advocated violence.

In sentence A, "they" refers to the councilmen, while in B, it refers to the demonstrators. The Winograd Schema Challenge, a benchmark for this task, was declared solved in 2019 after an LLM achieved 90 percent accuracy.

WIREHEADING

The act of electrically stimulating the brain's reward center to induce pleasure. In the 1950s, wireheaded rats were shown to forgo all other behavior in favor of self-stimulating in this way.

WORLD MODEL

A low-dimensional, stable representation of reality that captures essential structures and relationships, as opposed to a complex web of millions of statistical associations.

X

XOR OPERATION

A logical operation that fires only if one of its two inputs is switched on.

Z

ZERO-SHOT

Impromptu; when a model is prompted to perform a task without being given examples of successful performance. LLMs are weakest under these conditions.

Endnotes

PREFACE

1   Doyle, "The Red-Headed League."
2   van der Weij et al., "AI Sandbagging."
3   Bellard, "Text Compression."
4   Andrej Karpathy, "Jagged Intelligence,"
     X, July 25, 2024, https://x.com
     /karpathy/status/1816531576228053133;
     Dell'Acqua et al., "Jagged Technological
     Frontier"; Zhang et al., "Self-Contrast";
     Nezhurina et al., "Alice in Wonderland."
5   Bender et al., "Dangers of Stochastic Parrots."
6   Notopoulos, "A Chevy Dealership Added an AI
     Chatbot to Its Site."
7   Milmo, "Impossible to Create AI Tools Like
     ChatGPT without Copyrighted Material";
     Patronus AI, "Introducing CopyrightCatcher";
     Gold, "French Regulator Fines Google $271M";
     Helms and Kreiser, "Copyright Chaos."
8   Patnaik, Reinicke, and Bloomberg,
     "Nvidia Dethroned as World's Most
     Valuable Company."
9   Halperin, "AGI and the EMH."
10   Efrati and Holmes, "OpenAI Could Lose $5
     Billion This Year."
11   Brody (@heshiebee), "Google's AI overview
     recommends eating rocks," X, May 24, 2024,
     https://x.com/heshiebee/status
     /1793810016199197097.
12   Bilton, "Artificial Intelligence."
13   Future of Life Institute, "Pause Giant AI
     Experiments."
14   Shah et al., "Black-Box Jailbreaks"; Wei et al.,
     "Jailbroken."
15   Roose, "Conversation with Bing's Chatbot";
     Perrigo, "Bing's AI Is Threatening Users."
16   Perez, "Character.ai Is Catching Up to ChatGPT."
17   Verma, "They Fell in Love with AI Bots."
18   Semrush, "Most Visited Websites."

CHAPTER 1

19   Epoch AI, "Machine Learning Trends."
20   Moore, "Moore's Law at 40."
21   Epoch AI, "Cost to Train Frontier AI Models";
     Hestness et al., "Scaling Is Predictable"; Kaplan
     et al., "Scaling Laws."
22   For an extremely clear and early treatment of
     the relationship between compute, data, pa-
     rameters, and their ratios, see Nostalgebraist,
     "Scaling 'Inconsistency.'"
23   Kaplan et al., "Scaling Laws"; Hoffmann et al.,
     "Training Compute-Optimal Large Language
     Models."
24   Shulman and Bostrom, "How Hard Is Artificial
     Intelligence?"
25   Herculano-Houzel, "Human Brain as a Scaled-
     Up Primate Brain."
26   Kremer, "Population Growth and Technological
     Change."
27   Kaplan et al., "Scaling Laws." Brown et al.'s
     "Language Models Are Few-Shot Learners"
     famously demonstrates this trend for few-shot
     learning.

28   Jones, "Scaling Scaling Laws."
29   Ouyang et al., "Training Language Models to
     Follow Instructions."
30   Gemini Team, "Gemini."
31   Aschenbrenner, "Situational Awareness."
32   See, for example, Fu, Peng, and Khot, "How
     Does GPT Obtain Its Ability?"
33   Cireşan, Meier, and Schmidhuber, "Deep
     Neural Networks for Image Classification";
     Krizhevsky et al., "ImageNet Classification."
34   See, for example, Nostalgebraist, "GPT-3: A
     Disappointing Paper."
35   "The Scaling Hypothesis," excerpted in the
     Appendix.
36   Hestness et al., "Scaling Is Predictable."
37   See, for example, Medawar, "Is the Scientific
     Paper a Fraud?"
38   Branwen, "GANs Didn't Fail."
39   See, for example, Byrnes, "Brain-Like AGI
     Safety."
40   See Sperry et al., "Verbal Environments of
     Children."

CHAPTER 2

41   An up-to-date discussion of the difference is
     Schaeffer et al., "Downstream Capabilities of
     Frontier AI Models."
42   Shlegeris et al., "Next-Token Prediction."
43   See, for example, Constantin, "General
     Intelligences," and Moulton, "Digital Minds."
44   For details of these three theories, see Dziri et al.,
     "Faith and Fate," as well as the interviews with
     Shane Legg and Gwern Branwen in this book.
45   Deutsch, "Church-Turing Thesis."
46   Protzko and Colom, "Structure of Human
     Cognitive Ability."
47   Gemini Team, "Gemini 1.5."
48   See the ARC Prize,
     https://arcprize.org/.
49   Yudkowsky, "My Childhood Role Model."
50   Zhang et al., "Large Language Model
     Performance on Grade School Arithmetic."

CHAPTER 3

51   Vafa et al., "World Model Implicit in a
     Generative Model."
52   Recht et al., "ImageNet Classifiers"; Zhang et
     al., "Transcendence."
53   Srivastava et al., "Reasoning Performance."
54   Goldstein and Levinstein, "Does ChatGPT
     Have a Mind?"
55   Templeton et al., "Scaling Monosemanticity."
56   Rogers et al., "Is Autism a Disease of the
     Cerebellum?"
57   See Knolle, "Knowing What's Next."
58   Kanerva, Sparse Distributed Memory.
59   Elhage et al., "Toy Models of Superposition."
60   See, for example, Templeton et al., "Scaling
     Monosemanticity."
61   See, for example, Olshausen and Field,
     "Sparse Coding."

62   Cotra, "AI Takeover."
63   See, for example, Wei et al., "Jailbroken," and Qi et al., "Safety Alignment."

CHAPTER 4

64   For a discussion of the automated jailbreaking of even the most "aligned" models, see Wei et al., "Jailbroken," and Chao et al., "Jailbreaking Black Box Large Language Models." For an entertaining account of new jailbreaks, see Pliny the Prompter (@elder_plinius on X).
65   Branwen, "Tool AIs Want to Be Agent AIs." For a contemporary take on the same theme, see Kulveit et al., "Predictive Minds."
66   Nakano et al., "Browser-Assisted Question-Answering"; Adept Team, "ACT-1."
67   Bengio, "AI Safety."
68   Shlegeris et al., "Next-Token Prediction."
69   Transhumanist Ben (@amtrpa), "Consciousness is when the mask eats the shoggoth," X, March 6, 2023, https://x.com/amtrpa/status/1632737641941594116.
70   Hubinger et al., "Sleeper Agents."
71   Stromberg, "Neuroscientist Who Discovered He Was a Psychopath."
72   Anthropic, "Responsible Scaling Policy"; DeepMind, "Frontier Safety Framework"; OpenAI, "Preparedness Framework."
73   Maruf, "Google Fires Engineer Who Contended Its AI Technology Was Sentient."

CHAPTER 5

74   Patel, Nishball, and Ontiveros, "Energy Dilemma"; Morales, "World's Fastest Data Center"; Lyons, "Data Center Energy."
75   EPRI, "Powering Intelligence."
76   Epoch AI, "Notable AI Models."
77   OpenAI, "Partnership with News Corp" and "Reddit Partnership."
78   Abergotti and Matsakis, "OpenAI Has Hired an Army of Contractors"; Sang-deok and Deok-ju, "오픈AI, 물리학 난제 도전 범용과학 인공지능 만든다."
79   Dzieza, "Inside the AI Factory."
80   Patel and Wong, "GPT-4 Architecture, Infrastructure, Training Dataset."
81   Miller, "Gigawatt Data Center Campus"; Chernicoff, "Newly Acquired Nuclear Power Data Center."
82   Gardizy and Efrati, "$100 Billion Stargate AI Computer."
83   UNECE, "Labour Share of GDP."
84   Nuclear Newswire, "Amazon Buys Nuclear-Powered Data Center." However, the effort was blocked in November 2024. See Denning, "Amazon's Nuclear Deal Stalled."
85   See Hogarth, "Slow Down the Race to God-Like AI."
86   Villalobos et al., "Will We Run Out of ML Data?"

CHAPTER 6

87   This quote is attributed to von Neumann in Dyson, *Infinite in All Directions*.
88   See, for example, Ringel Morris et al., "Operationalizing Progress"; Schaeffer et al., "Downstream Capabilities of Frontier AI Models"; Leech et al., "Methods Failing the Data."
89   *The Economist*, "What Happened to the Artificial Intelligence Revolution?"; ZoomRx, "State of AI Report"; Mok, "Companies That Have Restricted Employees from Using ChatGPT."
90   Halperin et al., "AGI and the EMH."
91   As in Rogers, "Diffusion of Innovations."
92   Wang, "Transformative AI."
93   Attributed to the futurist Roy Amara. See Ratcliffe, "Roy Amara."
94   Chandra and Tabachnyk, "AI in Software Engineering at Google."
95   Forristal, "Duolingo Cuts 10% of Its Contractor Workforce"; Gerken, "AI Lets Us Cut Thousands of Jobs."
96   Dell'Acqua et al., "Jagged Technological Frontier."
97   Ibrahim et al., "Rethinking Homework"; Caplan, "GPT Retakes My Midterm"; Stribling et al., "Model Student."
98   Walton Family Foundation, "AI in Today's Classrooms."
99   See Cowen and Southwood, "Is the Rate of Scientific Progress Slowing Down?" and Bloom et al., "Are Ideas Getting Harder to Find?"
100   Fung, "Data Center IT Capex."
101   Kenwood, "Railway Investment in Britain."
102   US Senate, "Testimony of Dario Amodei."
103   Anthropic, "Frontier Threats Red Teaming."
104   Askell, "General Language Assistant."

CHAPTER 7

105   Fu et al., "GPTScore"; Vicuna Team, "Vicuna"; Zheng et al., "LLM-as-a-Judge"; van Schaik and Pugh, "Evaluation of LLM-Generated Summaries."
106   Branwen, "Slowing Moore's Law."
107   Bloom et al., "Are Ideas Getting Harder to Find?"
108   Besiroglu, "Are Models Getting Harder to Find?"
109   Hobbhahn, Heim, and Aydos, "Trends in Machine Learning Hardware."
110   Carmack, "Code for AGI Will Be Simple."
111   Karnofsky, "Most Important Century."

CHAPTER 8

112   Ryan Moulton (@moultano), "When our most advanced tech was fire, people thought life was like fire, and they were right," X, March 12, 2023, https://x.com/moultano/status/1634947952811872258.

113 See his April Fools' Day post. Yudkowsky, "MIRI Announces New 'Death with Dignity' Strategy."

114 For more on this analogy, see Yudkowsky, "Rocket Alignment Problem."

CONCLUSION

115 Alexander, "AI Size Solves Flubs"; JakubK, "GPT-4 Solves Gary Marcus-Induced Flubs"; Shengwu Li (@ShengwuLi), "It took me 30 seconds to fact-check this claim in the Chomsky op-ed," X, March 8, 2023, https://x.com/ShengwuLi/status/1633585663504261120; William MacAskill (@willmacaskill), "As an aside, it seems that this was true of Pinker, too," X, November 25, 2023, https://x.com/willmacaskill/status/1728486657366880452.

116 As expressed in Amodei's "Machines of Loving Grace," Altman's "Intelligence Age," and Karnofsky's "Forecasting Transformative AI."

PROFILES

117 Brown et al., "Language Models Are Few-Shot Learners."

118 Amodei et al., "Problems in AI Safety"; Amodei et al., "Deep Speech 2."

119 Aschenbrenner, "Situational Awareness."

120 Branwen, "Scaling Hypothesis."

121 Bricken and Pehlevan, "Attention"; Bricken et al., "Sparse Representations."

122 Bricken et al., "Towards Monosemanticity."

123 Carlsmith, "Is Power-Seeking AI an Existential Risk?"

124 Carlsmith, "Otherness and Control in the Age of AGI."

125 Chollet et al., "Xception."

126 Cotra, "Draft Report on AI Timelines."

127 Gemini Team, "Gemini"; Gemma Team, "Gemma"; Pope et al., "Scaling Transformer Inference."

128 Silver et al., "Mastering the Game of Go"; Silver et al., "Mastering Chess and Shogi"; Jumper et al., "Protein Structure Prediction."

129 Chen et al., "Decision Transformer."

130 Kaplan et al., "Scaling Laws."

131 Bai et al., "Constitutional AI"; Ganguli et al., "Red Teaming."

132 Karnofsky, "Hits-Based Giving"; Karnofsky, "Most Important Century."

133 Hutter, "Universal Intelligence."

134 Goertzel, "Who Coined the Term 'AGI'?"

135 Mnih et al., "Deep Reinforcement Learning"; Christiano et al., "Deep Reinforcement Learning."

136 Patel and Nishball, "Google Gemini Eats the World."

137 Nakano et al., "WebGPT."

138 Schulman, "Optimizing Expectations"; Schulman et al., "Proximal Policy Optimization Algorithms."

139 Krizhevsky et al., "ImageNet Classification."

140 Sutskever et al., "Sequence to Sequence Learning."

141 Eliezer Yudkowsky (@ESYudkowsky), "It is irrelevant how much AI labs spend on alignment because they have no hope of solving the problem," X, August 1, 2024, https://x.com/ESYudkowsky/status/1818995908790624646; Yudkowsky, "Pausing AI Developments Isn't Enough."

142 Meta, "Introducing Meta Llama 3."

Bibliography

A

Adept Team. "ACT-1: Transformer for Actions." Adept, September 14, 2022. https://www.adept.ai/blog/act-1.

Ahn, Michael, Anthony Brohan, Noah Brown, Yevgen Chebotar, Omar Cortes, Byron David, Chelsea Finn, Chuyuan Fu, Keerthana Gopalakrishnan, Karol Hausman, et al. "Do As I Can, Not As I Say: Grounding Language in Robotic Affordances." GitHub. Last updated August 16, 2022. https://say-can.github.io/.

Akyürek, Ekin, Dale Schuurmans, Jacob Andreas, Tengyu Ma, and Denny Zhou. "What Learning Algorithm Is In-Context Learning? Investigations with Linear Models." arXiv, November 28, 2022. https://arxiv.org/abs/2211.15661.

Alexander, Scott. "My Bet: AI Size Solves Flubs." Astral Codex Ten, June 7, 2022. https://www.astralcodexten.com/p/my-bet-ai-size-solves-flubs.

AlphaProof and AlphaGeometry Teams. "AI Achieves Silver-Medal Standard Solving International Mathematical Olympiad Problems." DeepMind, July 25, 2024. https://deepmind.google/discover/blog/ai-solves-imo-problems-at-silver-medal-level/.

Altman, Sam. "The Intelligence Age." September 23, 2024. https://ia.samaltman.com/.

Amodei, Dario. "Claude, New Models, AI Safety, and Economic Impact." *In Good Company*, June 26, 2024. Podcast, 67 min., 14 sec. https://www.youtube.com/watch?v=xm6jNMSFT7g.

Amodei, Dario. "Machines of Loving Grace: How AI Could Transform the World for the Better." October 2024. https://darioamodei.com/machines-of-loving-grace.

Amodei, Dario. "Written Testimony of Dario Amodei, PhD, Co-Founder and CEO, Anthropic." Hearing on Oversight of AI: Principles for Regulation Before the Judiciary Committee Subcommittee on Privacy, Technology, and the Law. United States Senate, July 25, 2023. https://www.judiciary.senate.gov/imo/media/doc/2023-07-26_-_testimony_-_amodei.pdf.

Amodei, Dario, Rishita Anubhai, Eric Battenberg, Carl Case, Jared Casper, Bryan Catanzaro, Jingdong Chen, Mike Chrzanowski, Adam Coates, Greg Diamos, et al. "Deep Speech 2: End-to-End Speech Recognition in English and Mandarin." arXiv, December 8, 2015. https://arxiv.org/abs/1512.02595.

Amodei, Dario, and Danny Hernandez. "AI and Compute." OpenAI, May 16, 2018. https://openai.com/index/ai-and-compute/.

Amodei, Dario, Chris Olah, Jacob Steinhardt, Paul Christiano, John Schulman, and Dan Mané. "Concrete Problems in AI Safety." arXiv, June 21, 2016. https://arxiv.org/abs/1606.06565.

Anderson, James A., and Edward Rosenfeld, eds. *Talking Nets: An Oral History of Neural Networks*. MIT Press, 2000.

Anthropic. "Core Views on AI Safety: When, Why, What, and How." March 8, 2023. https://www.anthropic.com/news/core-views-on-ai-safety.

Anthropic. "Frontier Threats Red Teaming for AI Safety." July 26, 2023. https://www.anthropic.com/news/frontier-threats-red-teaming-for-ai-safety.

Anthropic. "Responsible Scaling Policy." October 15, 2024. https://assets.anthropic.com/m/24a47b00f10301cd/original/Anthropic-Responsible-Scaling-Policy-2024-10-15.pdf.

Armstrong, Stuart, and Anders Sandberg. "Eternity in Six Hours: Intergalactic Spreading of Intelligent Life and Sharpening the Fermi Paradox." *Acta Astronautica* 89 (2013): 1–13. https://doi.org/10.1016/j.actaastro.2013.04.002.

Aschenbrenner, Leopold. "Situational Awareness: The Decade Ahead." Situational Awareness, June 2024. https://situational-awareness.ai/.

Askell, Amanda, Yuntao Bai, Anna Chen, Dawn Drain, Deep Ganguli, Tom Henighan, Andy Jones, Nicholas Joseph, Ben Mann, Nova DasSarma, et al. "A General Language Assistant as a Laboratory for Alignment." arXiv, December 1, 2021. https://arxiv.org/abs/2112.00861.

Austin, Jacob, Augustus Odena, Maxwell Nye, Maarten Bosma, Henryk Michalewski, David Dohan, Ellen Jiang, Carrie Cai, Michael Terry, Quoc Le, et al. "Program Synthesis with Large Language Models." arXiv, August 16, 2021. https://arxiv.org/abs/2108.07732.

"Average Nvidia GeForce RTX 3090 24GB Price History." HowMuch.one. https://howmuch.one/product/average-NVIDIA-geforce-rtx-3090-24gb/price-history.

B

Bahdanau, Dzmitry, Kyunghyun Cho, and Yoshua Bengio. "Neural Machine Translation by Jointly Learning to Align and Translate." arXiv, September 1, 2014. https://arxiv.org/abs/1409.0473.

Bai, Yuntao, Saurav Kadavath, Sandipan Kundu, Amanda Askell, Jackson Kernion, Andy Jones, Anna Chen, Anna Goldie, Azalia Mirhoseini, Cameron McKinnon, et al. "Constitutional AI: Harmlessness from AI Feedback." arXiv, December 15, 2022. https://arxiv.org/abs/2212.08073.

Barak, Boaz. "Emergent Abilities and Grokking: Fundamental, Mirage, or Both?" Windows on Theory, December 22, 2023. https://windowsontheory.org/2023/12/22/emergent-abilities-and-grokking-fundamental-mirage-or-both/.

Barnett, Matthew, and Tamay Besiroglu. "The Direct Approach." Epoch AI, April 25, 2023. https://epochai.org/blog/the-direct-approach.

Barnett, Peter. "Understanding Gradient Hacking." AI Alignment Forum, December 10, 2021. https://www.alignmentforum.org/posts/bdayaswyewjxxrQmB/understanding-gradient-hacking.

Belkin, Mikhail, Daniel Hsu, Siyuan Ma, and Soumik Mandal. "Reconciling Modern Machine Learning Practice and the Bias-Variance Trade-Off." arXiv, December 28, 2018. https://arxiv.org/abs/1812.11118.

Bell, Robert M., Yehuda Koren, and Chris Volinsky. "The BellKor 2008 Solution to the Netflix Prize." https://cseweb.ucsd.edu/classes/fa17/cse291-b/reading/ProgressPrize2008_BellKor.pdf.

Bellard, Fabrice. "Text Compression Using Large Language Models." 2023. https://bellard.org/ts_zip/.

Benaich, Nathan. "State of AI Report Compute Index." Air Street Capital. Last updated November 2024. https://www.stateof.ai/compute.

Bender, Emily M., Timnit Gebru, Angelina McMillan-Major, and Margaret Mitchell. "On the Dangers of Stochastic Parrots: Can Language Models Be Too Big?" *FAccT '21: Proceedings of the 2021 ACM Conference on Fairness, Accountability, and Transparency* (2021): 610–23. https://dl.acm.org/doi/10.1145/3442188.3445922.

Bengio, Yoshua. "Reasoning through Arguments Against Taking AI Safety Seriously." July 9, 2024. https://yoshuabengio.org/2024/07/09/reasoning-through-arguments-against-taking-ai-safety-seriously/.

Berglund, Lukas, Meg Tong, Max Kaufmann, Mikita Balesni, Asa Cooper Stickland, Tomasz Korbak, and Owain Evans. "The Reversal Curse: LLMs Trained on 'A Is B' Fail to Learn 'B Is A.'" arXiv, September 21, 2023. https://arxiv.org/abs/2309.12288.

Besiroglu, Tamay. "Are Models Getting Harder to Find?" MPhil thesis, University of Cambridge, August 2020. https://github.com/Besiroglu/webpage/blob/3682ccac6fc92378934c24b0c08a64bcca1793e6/papers/AreModels.pdf.

Betker, James. "The 'It' in AI Models Is the Dataset." Non_Interactive, June 10, 2023. https://nonint.com/2023/06/10/the-it-in-ai-models-is-the-dataset/.

Bilton, Nick. "Artificial Intelligence May Be Humanity's Most Ingenious Invention—And Its Last?" *Vanity Fair*, September 13, 2023. https://www.vanityfair.com/news/2023/09/artificial-intelligence-industry-future.

Bloom, Nicholas, Charles I. Jones, John Van Reenen, and Michael Webb. "Are Ideas Getting Harder to Find?" *American Economic Review*, 110, no. 4 (2020): 1104–44. https://www.aeaweb.org/articles?id=10.1257/aer.20180338.

Bostrom, Nick. "Ethical Issues in Advanced Artificial Intelligence." https://nickbostrom.com/ethics/ai.

Bostrom, Nick, and Milan Ćirković, eds. *Global Catastrophic Risks*. Oxford University Press, 2011.

Bowman, Samuel R., Jeeyoon Hyun, Ethan Perez, Edwin Chen, Craig Pettit, Scott Heiner, Kamilė Lukošiūtė, Amanda Askell, Andy Jones, Anna Chen, et al. "Measuring Progress on Scalable Oversight for Large Language Models." arXiv, November 4, 2022. https://arxiv.org/abs/2211.03540.

Bradbury, Ray. "Night Call, Collect." In *I Sing the Body Electric! Stories by Ray Bradbury*. Random House, 1969.

Brants, Thorsten, Ashok C. Popat, Peng Xu, Franz J. Och, and Jeffrey Dean. "Large Language Models in Machine Translation." *Proceedings of the 2007 Joint Conference on Empirical Methods in Natural Language Processing and Computational Natural Language Learning* (2007): 858–67. https://aclanthology.org/D07-1090.pdf.

Branwen, Gwern. "Comment on 'Bing Chat Is Blatantly, Aggressively Misaligned.'" LessWrong, February 17, 2023. https://www.lesswrong.com/posts/jtoPawEhLNXNxvgTT/bing-chat-is-blatantly-aggressively-misaligned#AAC8jKeDp6xqsZK2K.

Branwen, Gwern. "Comment on 'If I Wanted to Spend Way More on AI, What Would I Spend It On?'" LessWrong, September 16, 2024. https://www.lesswrong.com/posts/bX7q9NcoGpb5KdZzQ/if-i-wanted-to-spend-way-more-on-ai-what-would-i-spend-it-on#rWDffqwcaCiMfpTTt.

Branwen, Gwern. "GANs Didn't Fail, They Were Abandoned." Gwern.net, October 4, 2022. https://gwern.net/gan.

Branwen, Gwern. "The Scaling Hypothesis." Gwern.net, May 28, 2020. https://gwern.net/scaling-hypothesis.

Branwen, Gwern. "Slowing Moore's Law: How It Could Happen." Gwern.net, March 16, 2012. https://gwern.net/slowing-moores-law.

Branwen, Gwern. "Why Tool AIs Want to Be Agent AIs." Gwern.net, September 7, 2016. https://gwern.net/tool-ai.

Bricken, Trenton, and Cengiz Pehlevan. "Attention Approximates Sparse Distributed Memory." arXiv, November 10, 2021. https://arxiv.org/abs/2111.05498.

Bricken, Trenton, Rylan Schaeffer, Bruno Olshausen, and Gabriel Kreiman. "Emergence of Sparse Representations from Noise." *Proceedings of the 40th International Conference on Machine Learning* (2023). https://proceedings.mlr.press/v202/bricken23a/bricken23a.pdf.

Bricken, Trenton, Adly Templeton, Joshua Batson, Brian Chen, Adam Jermyn, Tom Conerly, Nicholas L. Turner, Cem Anil, Carson Denison, Amanda Askell, et al. "Towards Monosemanticity: Decomposing Language Models With Dictionary Learning." Transformer Circuits Thread, October 4, 2023. https://transformer-circuits.pub/2023/monosemantic-features.

Brown, Bradley, Jordan Juravsky, Ryan Ehrlich, Ronald Clark, Quoc V. Le, Christopher Ré, and Azalia Mirhoseini. "Large Language Monkeys: Scaling Inference Compute with Repeated Sampling." arXiv, July 31, 2024. https://arxiv.org/abs/2407.21787.

Brown, Tom B., Benjamin Mann, Nick Ryder, Melanie Subbiah, Jared Kaplan, Prafulla Dhariwal, Arvind Neelakantan, Pranav Shyam, Girish Sastry, Amanda Askell, et al. "Language Models Are Few-Shot Learners." arXiv, May 28, 2020. https://arxiv.org/abs/2005.14165.

Bubeck, Sebastien, Varun Chandrasekaran, Ronen Eldan, Johannes Gehrke, Eric Horvitz, Ece Kamar, Peter Lee, Yin Tat Lee, Yuanzhi Li, Scott Lundberg, et al. "Sparks of Artificial General Intelligence: Early Experiments with GPT-4." arXiv, March 22, 2023. https://arxiv.org/abs/2303.12712.

Burns, Collin, Pavel Izmailov, Jan Hendrik Kirchner, Bowen Baker, Leo Gao, Leopold Aschenbrenner, Yining Chen, Adrien Ecoffet, Manas Joglekar, Jan

Leike, et al. "Weak-to-Strong Generalization: Eliciting Strong Capabilities With Weak Supervision." arXiv, December 14, 2023. https://arxiv.org/abs/2312.09390.

Byrnes, Steven. "Intro to Brain-Like-AGI Safety." LessWrong, January 26, 2022. https://www.lesswrong.com/s/HzcM2dkCq7fwXBej8.

C

"Caesar Cipher Encoding Request." ChatGPT, October 14, 2024. https://chatgpt.com/share/670d42e3-7edc-8009-9007-6f32ac1f4358.

Cahn, David. "AI's $600B Question." Sequoia, June 20, 2024. https://www.sequoiacap.com/article/ais-600b-question/.

Campos, Daniel. "Curriculum Learning for Language Modeling." arXiv, August 4, 2021. https://arxiv.org/abs/2108.02170.

Caplan, Bryan. "GPT Retakes My Midterm and Gets an AI." Bet on It, March 21, 2023. https://www.betonit.ai/p/gpt-retakes-my-midterm-and-gets-an.

Carchidi, Vincent J. "Is OpenAI's Sam Altman's Future Worth $7 Trillion?" The Hill, April 7, 2024. https://thehill.com/opinion/4579411-is-openais-sam-altmans-future-worth-7-trillion/.

Carlsmith, Joseph. "Is Power-Seeking AI an Existential Risk?" arXiv, June 16, 2022. https://arxiv.org/abs/2206.13353.

Carlsmith, Joseph. "Otherness and Control in the Age of AGI." January 2, 2024. https://jc.gatspress.com/pdf/otherness_full.pdf.

Carmack, John. "The Code for AGI Will Be Simple." Lex Fridman Podcast, August 6, 2022. Video, 16 min., 02 sec. https://youtube.com/watch?v=xLi83prR5fg.

Cattell, R. B. "The Measurement of Adult Intelligence." Psychological Bulletin 40, no. 3 (1943): 153–93. https://doi.org/10.1037/h0059973.

Chandra, Satish, and Maxim Tabachnyk. "AI in Software Engineering at Google: Progress and the Path Ahead." Google Research, June 6, 2024. https://research.google/blog/ai-in-software-engineering-at-google-progress-and-the-path-ahead.

Chao, Patrick, Alexander Robey, Edgar Dobriban, Hamed Hassani, George J. Pappas, and Eric Wong. "Jailbreaking Black Box Large Language Models in Twenty Queries." arXiv, October 12, 2023. https://arxiv.org/abs/2310.08419.

Chen, Lili, Kevin Lu, Aravind Rajeswaran, Kimin Lee, Aditya Grover, Michael Laskin, Pieter Abbeel, Aravind Srinivas, and Igor Mordatch. "Decision Transformer: Reinforcement Learning via Sequence Modeling." arXiv, June 2, 2021. https://arxiv.org/abs/2106.01345.

Chen, Mark, Jerry Tworek, Heewoo Jun, Qiming Yuan, Henrique Ponde de Oliveira Pinto, Jared Kaplan, Harri Edwards, Yuri Burda, Nicholas Joseph, Greg Brockman, et al. "Evaluating Large Language Models Trained on Code." arXiv, July 7, 2021. https://arxiv.org/abs/2107.03374.

Chen, Xiangning, Chen Liang, Da Huang, Esteban Real, Kaiyuan Wang, Yao Liu, Hieu Pham,

Xuanyi Dong, Thang Luong, Cho-Jui Hsieh, et al. "Symbolic Discovery of Optimization Algorithms." arXiv, May 8, 2023. https://arxiv.org/abs/2302.06675.

Chen, Yutian, Aja Huang, Ziyu Wang, Ioannis Antonoglou, Julian Schrittwieser, David Silver, and Nando de Freitas. "Bayesian Optimization in AlphaGo." arXiv, December 17, 2018. https://arxiv.org/abs/1812.06855.

Chernicoff, David. "AWS Eyes 960 MW for Newly Acquired Nuclear Power Data Center in Pennsylvania." Data Center Frontier, March 6, 2024. https://www.datacenterfrontier.com/hyperscale/article/33038288/aws-eyes-960-mw-for-newly-acquired-nuclear-power-data-center-in-pennsylvania.

Chiang, Ted. "ChatGPT Is a Blurry JPEG of the Web." The New Yorker, February 9, 2023. https://www.newyorker.com/tech/annals-of-technology/chatgpt-is-a-blurry-jpeg-of-the-web.

Chollet, François. "Xception: Deep Learning With Depthwise Separable Convolutions." Proceedings of the IEEE Conference on Computer Vision and Pattern Recognition (2017): 1251–58. https://openaccess.thecvf.com/content_cvpr_2017/html/Chollet_Xception_Deep_Learning_CVPR_2017_paper.html.

Chollet, François, Katherine Tong, Walter Reade, and Julia Elliott. "Abstraction and Reasoning Challenge." Kaggle, February 13, 2020. https://kaggle.com/competitions/abstraction-and-reasoning-challenge.

Christiano, Paul. "Clarifying 'AI Alignment.'" AI Alignment, April 7, 2018. https://ai-alignment.com/clarifying-ai-alignment-cec47cd69dd6.

Christiano, Paul F., Jan Leike, Tom Brown, Miljan Martic, Shane Legg, and Dario Amodei. "Deep Reinforcement Learning from Human Preferences." In Advances in Neural Information Processing Systems 30 (NIPS 2017), edited by I. Guyon, U. von Luxburg, S. Bengio, H. Wallach, R. Fergus, S. Vishwanathan, and R. Garnett. 2017. https://proceedings.neurips.cc/paper_files/paper/2017/hash/d5e2c0adad503c91f91df240d0cd4e49-Abstract.html.

Cimatti, Alessandro, Marco Pistore, and Paolo Traverso. "Chapter 22 Automated Planning." Foundations of Artificial Intelligence 3 (2008): 841–67. https://doi.org/10.1016/S1574-6526(07)03022-2.

Cireşan, Dan, Ueli Meier, and Juergen Schmidhuber. "Multi-column Deep Neural Networks for Image Classification." arXiv, February 13, 2012. https://arxiv.org/abs/1202.2745.

Cohen, Avner, and William Burr, eds. "The U.S. Discovery of Israel's Secret Nuclear Project." National Security Archive Electronic Briefing Book No. 510. April 15, 2015. https://nsarchive2.gwu.edu/nukevault/ebb510/.

Constantin, Sarah R. "Humans Who Are Not Concentrating Are Not General Intelligences." February 25, 2019. https://srconstantin.wordpress.com/2019/02/25/humans-who-are-not-concentrating-are-not-general-intelligences/.

Cook, Robert. "Kohler's Research on the Mentality of Apes." Tufts University, Spring Semester 2013. https://pigeon.psy.tufts.edu/psych26/kohler.htm.

Cotra, Ajeya. "Draft Report on AI Timelines." LessWrong, September 19, 2020. https://www.lesswrong.com/posts/KrJfoZzpSDpnrv9va/draft-report-on-ai-timelines.

Cotra, Ajeya. "Scale, Schlep, and Systems." Planned Obsolescence, October 10, 2023. https://www.planned-obsolescence.org/scale-schlep-and-systems/.

Coupé, Christophe, Yoon Mi Oh, Dan Ddiu, and François Pellegrino. "Different Languages, Similar Encoding Efficiency: Comparable Information Rates Across the Human Communicative Niche." *Science Advances* 5, no. 9 (2019). https://doi.org/10.1126/sciadv.aaw2594.

Cowen, Tyler, and Ben Southwood. "Is the Rate of Scientific Progress Slowing Down?" GMU Working Paper in Economics No. 21-13, August 5, 2019. https://papers.ssrn.com/sol3/papers.cfm?abstract_id=3822691.

Crownhart, Casey. "Why Microsoft Made a Deal to Help Restart Three Mile Island." *MIT Technology Review*, September 26, 2024. https://www.technologyreview.com/2024/09/26/1104516/three-mile-island-microsoft/.

D

Dalrymple, David A. "A List of Core AI Safety Problems and How I Hope to Solve Them." AI Alignment Forum, August 26, 2023. https://www.alignmentforum.org/posts/mnoc3cKY3gXMrTybs/a-list-of-core-ai-safety-problems-and-how-i-hope-to-solve.

Deacon, Terrence W. *The Symbolic Species: The Co-Evolution of Language and the Brain*. W.W. Norton & Company, 1998.

Dell'Acqua, Fabrizio, Edward McFowland III, Ethan R. Mollick, Hila Lifshitz-Assaf, Katherine Kellogg, Saran Rajendran, Lisa Krayer, François Candelon, and Karim R. Lakhani. "Navigating the Jagged Technological Frontier." Harvard Business School Technology & Operations Mgt. Unit Working Paper No. 24-013, September 15, 2023. https://papers.ssrn.com/sol3/papers.cfm?abstract_id=4573321.

Denning, Liam. "Amazon's Nuclear Deal Stalled— But Its AI Power Demand Won't." Bloomberg, November 4, 2024. https://www.bloomberg.com/opinion/articles/2024-11-04/amazon-s-nuclear-deal-stalled-but-its-ai-power-demand-won-t.

Deutsch, David. "Quantum Theory, the Church–Turing Principle and the Universal Quantum Computer." *Proceedings of the Royal Society of London A* 400, no. 1818 (July 1985). https://royalsocietypublishing.org/doi/10.1098/rspa.1985.0070.

Ding, Jeffrey. *Technology and the Rise of Great Powers: How Diffusion Shapes Economic Competition*. Princeton University Press, 2024.

Doyle, Arthur Conan. "The Red-Headed League." In *The Penguin Complete Sherlock Holmes*. Penguin, 2009.

Dubey, Abhimanyu, Abhinav Jauhri, Abhinav Pandey, Abhishek Kadian, Ahmad Al-Dahle, Aiesha Letman, Akhil Mathur, Alan Schelten, Amy Yang, Angela Fan, et al. "The Llama 3 Herd of Models." arXiv, July 31, 2024. https://arxiv.org/abs/2407.21783.

Dyson, Freeman J. *Infinite in All Directions: Gifford Lectures Given at Aberdeen, Scotland, April–November 1985*. Harper & Row, 1988.

Dzieza, Josh. "Inside the AI Factory." The Verge, June 20, 2023. https://www.theverge.com/features/23764584/ai-artificial-intelligence-data-notation-labor-scale-surge-remotasks-openai-chatbots.

Dziri, Nouha, Ximing Lu, Melanie Sclar, Xiang Lorraine, Liwei Jiang, Bill Yuchen Lin, Peter West, Chandra Bhagavatula, Ronan Le Bras, Jena D. Hwang, et al. "Faith and Fate: Limits of Transformers on Compositionality." arXiv, May 29, 2023. https://arxiv.org/abs/2305.18654.

E

The Economist. "What Happened to the Artificial Intelligence Revolution?" July 2, 2024. https://www.economist.com/finance-and-economics/2024/07/02/what-happened-to-the-artificial-intelligence-revolution.

Edwards, Benj. "Anthropic's Claude 3 Causes Stir by Seeming to Realize When It's Being Tested." Ars Technica, April 5, 2024. https://arstechnica.com/information-technology/2024/03/claude-3-seems-to-detect-when-it-is-being-tested-sparking-ai-buzz-online/.

Efrati, Amir, and Aaron Holmes. "Why OpenAI Could Lose $5 Billion This Year." The Information, July 24, 2024. https://www.theinformation.com/articles/why-openai-could-lose-5-billion-this-year.

Electric Power Research Institute. "Powering Intelligence: Analyzing Artificial Intelligence and Data Center Energy Consumption." May 2024. https://www.epri.com/research/products/000000003002028905.

Elhage, Nelson. "Systems That Defy Detailed Understanding." February 22, 2020. https://blog.nelhage.com/post/systems-that-defy-understanding/.

Elhage, Nelson, Tristan Hume, Catherine Olsson, Nicholas Schiefer, Tom Henighan, Shauna Kravec, Zac Hatfield-Dodds, Robert Lasenby, Dawn Drain, Carol Chen, et al. "Toy Models of Superposition." Transformer Circuits Thread, September 14, 2022. https://transformer-circuits.pub/2022/toy_model/index.html.

Elhage, Nelson, Neel Nanda, Catherine Olsson, Tom Henighan, Nicholas Joseph, Ben Mann, Amanda Askell, Yuntao Bai, Anna Chen, Tom Conerly, et al. "A Mathematical Framework for Transformer Circuits." Transformer Circuits Thread, December 22, 2021. https://transformer-circuits.pub/2021/framework/index.html.

Epoch AI. "Data on Notable AI Models." June 19, 2024. https://epochai.org/data/notable-ai-models.

Epoch AI. "How Much Does It Cost to Train Frontier AI Models?" June 3, 2024. https://epochai.org/blog/how-much-does-it-cost-to-train-frontier-ai-models.

Epoch AI. "Machine Learning Trends." Last updated June 7, 2024. https://docs.google.com/document/d/1BZ-7cSD-BFFQdp86C-A742ALaZiTknD3mmPMXuyY0vw/edit.

Erdil, Ege, and Tamay Besiroglu. "Revisiting Algorithmic Progress." Epoch AI, December 12, 2022. https://epochai.org/blog/revisiting-algorithmic-progress.

Evans, Richard, and Jim Gao. "DeepMind AI Reduces Google Data Centre Cooling Bill by 40%." Google DeepMind, July 20, 2016. https://deepmind.google/discover/blog/deepmind-ai-reduces-google-data-centre-cooling-bill-by-40/.

F

Farquhar, Sebastian, Jannik Kossen, Lorenz Kuhn, and Yarin Gal. "Detecting Hallucinations in Large Language Models Using Semantic Entropy." *Nature* 630 (June 2024): 625–30. https://www.nature.com/articles/s41586-024-07421-0.

Faruqui, Manaal, Yulia Tsvetkov, Dani Yogatama, Chris Dyer, and Noah A. Smith. "Sparse Overcomplete Word Vector Representations." arXiv, June 5, 2015. https://arxiv.org/abs/1506.02004.

Fawzi, Alhussein, Matej Balog, Bernardino Romera-Paredes, Demis Hassabis, and Pushmeet Kohli. "Discovering Novel Algorithms with AlphaTensor." Google DeepMind, October 5, 2022. https://deepmind.google/discover/blog/discovering-novel-algorithms-with-alphatensor/.

FEMA. "The Defense Production Act of 1950, as Amended." August 13, 2018. https://www.fema.gov/sites/default/files/2020-03/Defense_Production_Act_2018.pdf.

Feng, Yunzhen, Elvis Dohmatob, Pu Yang, Francois Charton, and Julia Kempe. "Beyond Model Collapse: Scaling Up with Synthesized Data Requires Reinforcement." arXiv, June 11, 2024. https://arxiv.org/abs/2406.07515.

Fist, Tim, and Arnab Datta. "How to Build the Future of AI in the United States." Institute for Progress, October 3, 2024. https://ifp.org/future-of-ai-compute/.

Forristal, Lauren. "Duolingo Cuts 10% of Its Contractor Workforce as the Company Embraces AI." TechCrunch, January 9, 2024. https://techcrunch.com/2024/01/09/duolingo-cut-10-of-its-contractor-workforce-as-the-company-embraces-ai/.

Frey, Carl Benedikt, and Michael A. Osborne, "The Future of Employment: How Susceptible Are Jobs to Computerisation?" *Technological Forecasting and Social Change* 114 (2017): 254–80. https://doi.org/10.1016/j.techfore.2016.08.019.

Fu, Jinlan, See-Kiong Ng, Zhengbao Jiang, Pengfei Liu. "GPTScore: Evaluate as You Desire." arXiv, February 8, 2023. https://arxiv.org/pdf/2302.04166.

Fu, Yao, Hao Peng, and Tushar Koht. "How Does GPT Obtain Its Ability? Tracing Emergent Abilities of Language Models to Their Sources." Notion, December 11, 2022. https://yaofu.notion.site/How-does-GPT-Obtain-its-Ability-Tracing-Emergent-Abilities-of-Language-Models-to-their-Sources-b9a57ac0fcf74f30a1ab9e3e36fa1dc1.

Fung, Baron. "Market Research on Data Center IT Capex." Dell'Oro Group, 2024. https://www.delloro.com/market-research/data-center-infrastructure/data-center-capex/.

Future of Life Institute. "Pause Giant AI Experiments: An Open Letter." March 22, 2023. https://futureoflife.org/open-letter/pause-giant-ai-experiments/.

G

Gabrielsen, Cory. "The Waluigi Effect." cory.eth, February 22, 2023. https://coryeth.substack.com/p/the-waluigi-effect.

Ganguli, Deep, Liane Lovitt, Jackson Kernion, Amanda Askell, Yuntao Bai, Saurav Kadavath, Ben Mann, Ethan Perez, Nicholas Schiefer, Kamal Ndousse, et al. "Red Teaming Language Models to Reduce Harms: Methods, Scaling Behaviors, and Lessons Learned." arXiv, August 23, 2022. https://arxiv.org/abs/2209.07858.

Gardizy, Anissa, and Amir Efrati. "Microsoft and OpenAI Plot $100 Billion Stargate AI Computer." The Information, March 29, 2024. https://www.theinformation.com/articles/microsoft-and-openai-plot-100-billion-stargate-ai-supercomputer.

Gartenberg, Chaim. "What Is a Long Context Window?" Google, February 16, 2024. https://blog.google/technology/ai/long-context-window-ai-models.

Gemini Team. "Gemini: A Family of Highly Capable Multimodal Models." arXiv, December 19, 2023. https://arxiv.org/abs/2312.11805.

Gemini Team. "Gemini 1.5: Unlocking Multimodal Understanding across Millions of Tokens of Context." arXiv, March 8, 2024. https://arxiv.org/abs/2403.05530.

Gemma Team. "Gemma: Open Models Based on Gemini Research and Technology." arXiv, March 13, 2024. https://arxiv.org/abs/2403.08295.

Gerken, Tom. "Klarna: AI Lets Us Cut Thousands of Jobs—But Pay More." BBC, August 28, 2024. https://www.bbc.co.uk/news/articles/c80e1gp9m9zo.

Goertzel, Ben. "Who Coined the Term 'AGI'?" August 28, 2011. https://web.archive.org/web/20181228083048/http://goertzel.org/who-coined-the-term-agi/.

Gold, Jon. "French Regulator Fines Google $271M Over Generative AI Copyright Issue." CIO, March 20, 2024. https://www.cio.com/article/2069449/french-regulator-fines-google-271m-over-generative-ai-copyright-issue.html.

Goldie, Anna, Azalia Mirhoseini, Mustafa Yazgan, Joe Wenjie Jiang, Ebrahim Songhori, Shen Wang, Young-Joon Lee, Eric Johnson, Omkar Pathak, Azade Nova, et al. "Addendum: A Graph Placement Methodology for Fast Chip Design." *Nature* 634

(September 2024): E10–11. https://www.nature.com/articles/s41586-024-08032-5.

Goldstein, Simon, and Benjamin Anders Levinstein. "Does ChatGPT Have a Mind?" arXiv, June 27, 2024. https://arxiv.org/abs/2407.11015.

Good, Irving John. "Speculations Concerning the First Ultraintelligent Machine." *Advances in Computers* 6 (1966): 31–88. https://doi.org/10.1016/S0065-2458(08)60418-0.

Google DeepMind. "Frontier Safety Framework." May 17, 2024. https://storage.googleapis.com/deepmind-media/DeepMind.com/Blog/introducing-the-frontier-safety-framework/fsf-technical-report.pdf.

Goyal, Mohit, Rajan Goyal, and Brejesh Lall. "Learning Activation Functions: A New Paradigm for Understanding Neural Networks." arXiv, June 23, 2019. https://arxiv.org/abs/1906.09529.

Greenblatt, Ryan. "By When Will 85% Be Reached on the Public Evaluation Set on ARC-AGI-Pub?" Manifold. https://manifold.markets/RyanGreenblatt/by-when-will-85-be-reached-on-the-p.

Greenblatt, Ryan. "Getting 50% (SoTA) on ARC-AGI with GPT-4o." Redwood Research Blog, June 17, 2024. https://redwoodresearch.substack.com/p/getting-50-sota-on-arc-agi-with-gpt.

Grietzer, Peli. "The Problem With the Word 'Alignment.'" AI Objectives Institute. https://ai.objectives.institute/blog/the-problem-with-alignment.

Griffith, Erin. "The Desperate Hunt for the AI Boom's Most Indispensable Prize." *The New York Times*, August 16, 2023. https://www.nytimes.com/2023/08/16/technology/ai-gpu-chips-shortage.html.

H

Halevy, Alon, Peter Norvig, and Fernando Pereira. "The Unreasonable Effectiveness of Data." *IEEE Intelligent Systems* 24, no. 2 (March–April 2009): 8–12. https://ieeexplore.ieee.org/document/4804817.

Halperin, Basil, J. Zachary Mazlish, and Trevor Chow. "AGI and the EMH: Markets Are Not Expecting Aligned or Unaligned AI in the Next 30 Years." Effective Altruism Forum, January 10, 2023. https://forum.effectivealtruism.org/posts/8c7LycgtkypkgYjZx/agi-and-the-emh-markets-are-not-expecting-aligned-or.

Hambrick, David Z., Alexander P. Burgoyne, and Erik M. Altmann. "Problem-Solving and Intelligence." In *The Cambridge Handbook of Intelligence*. Cambridge University Press, 2019.

Hanson, Robin, and Eliezer Yudkowsky. *The Hanson-Yudkowsky AI-Foom Debate*. Machine Intelligence Research Institute, 2013. https://intelligence.org/files/AIFoomDebate.pdf.

Heffernan, Virginia. "I Saw the Face of God in a TSMC Factory." *Wired,* March 21, 2023. https://www.wired.com/story/i-saw-the-face-of-god-in-a-tsmc-factory/.

Heinlein, Robert A. *Stranger in a Strange Land*. Putnam, 1961.

Helms, Shawn C., and Jason D. Kreiser. "ChatGPT Will Unleash Copyright Chaos." *Barron's*, January 27, 2023. https://www.barrons.com/articles/chatgpt-will-unleash-copyright-chaos-artificial-intelligence-51674780407.

Hendrycks, Dan, Collin Burns, Saurav Kadavath, Akul Arora, Steven Basart, Eric Tang, Dawn Song, and Jacob Steinhardt. "Measuring Mathematical Problem Solving with the MATH Dataset." arXiv, March 5, 2021. https://arxiv.org/abs/2103.03874.

Herculano-Houzel, Suzana. "The Human Brain in Numbers: A Linearly Scaled-Up Primate Brain." *Frontiers in Human Neuroscience* 3, no. 31 (November 2009). 10.3389/neuro.09.031.2009.

Herculano-Houzel, Suzana. "The Remarkable, Yet Not Extraordinary, Human Brain as a Scaled-Up Primate Brain and Its Associated Cost." *PNAS* 109, suppl. 1 (2012): 10661–68. https://doi.org/10.1073/pnas.1201895109.

Hestness, Joel, Sharan Narang, Newsha Ardalani, Gregory Diamos, Heewoo Jun, Hassan Kianinejad, Md. Mostofa Ali Patwary, Yang Yang, and Yanqi Zhou. "Deep Learning Scaling Is Predictable, Empirically." arXiv, December 1, 2017. https://arxiv.org/abs/1712.00409.

Hinton, Geoffrey. "On Radiology." 2016 Machine Learning and Market for Intelligence Conference, Toronto, ON, November 2016. Video, 1 min., 24 sec. https://www.youtube.com/watch?v=2HMPRXstSvQ.

Hinton, Geoffrey, J.L. McClelland, and D.E. Rumelhart. "Distributed Representations." In *Parallel Distributed Processing: Explorations in the Microstructure of Cognition*. Association for Computing Machinery, 1986.

Hitzler, Pascal, Aaron Eberhart, Monireh Ebrahimi, Md Kamruzzaman Sarker, and Lu Zhou. "Neuro-Symbolic Approaches in Artificial Intelligence." *National Science Review* 9, no. 6 (2022): nwac035. https://doi.org/10.1093/nsr/nwac035.

Ho, Anson, Tamay Besiroglu, Ege Erdil, David Owen, Robi Rahman, Zifan Carl Guo, David Atkinson, Neil Thompson, and Jaime Sevilla. "Algorithmic Progress in Language Models." Epoch AI, March 12, 2024. https://epochai.org/blog/algorithmic-progress-in-language-models.

Hobart, Byrne, and Tobias Huber. "Manias and Mimesis: Applying René Girard's Mimetic Theory to Financial Bubbles." SSRN, October 11, 2019. http://dx.doi.org/10.2139/ssrn.3469465.

Hobbes, Thomas. *Leviathan or The Matter, Forme and Power of a Commonwealth Ecclesiasticall and Civil*. 1668.

Hobbhahn, Marius. "Understanding Strategic Deception and Deceptive Alignment." Apollo Research, September 25, 2024. https://www.apolloresearch.ai/blog/understanding-strategic-deception-and-deceptive-alignment.

Hobbhahn, Marius, and Tamay Besiroglu. "Trends in GPU Price Performance." EpochAI, June 27, 2022. https://epochai.org/blog/trends-in-gpu-price-performance.

Hobbhahn, Marius, Lennart Heim, Gökçe Aydos. "Trends in Machine Learning Hardware." EpochAI, November 9, 2023. https://epochai.org/blog/trends-in-machine-learning-hardware.

Hoffmann, Jordan, Sebastian Borgeaud, Arthur
Mensch, Elena Buchatskaya, Trevor Cai, Eliza
Rutherford, Diego de Las Casas, Lisa Anne
Hendricks, Johannes Welbl, Aidan Clark.
"Training Compute-Optimal Large Language
Models." arXiv, March 29, 2022. https://arxiv.org/
abs/2203.15556.

Hogarth, Ian. "We Must Slow Down the Race to God-
Like AI." *Financial Times*, April 12, 2023. https://
www.ft.com/content/03895dc4-a3b7-481e-
95cc-336a524f2ac2.

Hsu, Feng-Hsiung. *Behind Deep Blue: Building
the Computer That Defeated the World Chess
Champion.* Princeton University Press, 2002.

Huang, Dan. "How Much Did AlphaGo Cost?"
Dansplaining, March 2018. https://www.yuzeh.
com/data/agz-cost.html.

Huang, Jie, Xinyun Chen, Swaroop Mishra, Huaixiu
Steven Zheng, Adams Wei Yu, Xinying Song, and
Denny Zhou. "Large Language Models Cannot
Self-Correct Reasoning Yet." arXiv, October 3,
2023. https://arxiv.org/abs/2310.01798.

Hubinger, Evan, Carson Denison, Jesse Mu, Mike
Lambert, Meg Tong, Monte MacDiarmid,
Tamera Lanham, Daniel M. Ziegler, Tim Maxwell,
Newton Cheng, et al. "Sleeper Agents: Training
Deceptive LLMs That Persist through Safety
Training." arXiv, January 10, 2024. https://arxiv.org/
abs/2401.05566.

Hutter, Marcus. "A Theory of Universal Artificial
Intelligence Based on Algorithmic Complexity."
arXiv, April 3, 2000. https://arxiv.org/abs/
cs/0004001.

I

Ibrahim, Hazem, Rohail Asim, Fareed Zaffar, Talal
Rahwan, and Yasir Zaki. "Rethinking Homework in
the Age of Artificial Intelligence." *IEEE Intelligent
Systems* 38, no. 2 (2023): 24–7. https://doi.
org/10.1109/MIS.2023.3255599.

J

JakubK. "GPT-4 Solves Gary Marcus-Induced
Flubs." LessWrong, March 17, 2023. https://www.
lesswrong.com/posts/cGbEtNbxACJpqoP4x/
gpt-4-solves-gary-marcus-induced-flubs.

Jimenez, Carlos E., John Yang, Alexander Wettig,
Shunyu Yao, Kexin Pei, Ofir Press, and Karthik
Narasimhan. "SWE-Bench: Can Language
Models Resolve Real-World GitHub Issues?"
arXiv, October 10, 2023. https://arxiv.org/
abs/2310.06770.

Jones, Andy L. "Scaling Scaling Laws with Board
Games." arXiv, April 7, 2021. https://arxiv.org/
abs/2104.03113.

Jumper, John, Richard Evans, Alexander Pritzel, Tim
Green, Michael Figurnov, Olaf Ronneberger,
Kathryn Tunyasuvunakool, Russ Bates, Augustin
Žídek, Anna Potapenko, et al. "Highly Accurate
Protein Structure Prediction with AlphaFold."
Nature 596 (2021): 583–89. https://doi.
org/10.1038/s41586-021-03819-2.

K

Kanerva, Pentti. *Sparse Distributed Memory.* MIT
Press, 2003.

Kaplan, Jared, Sam McCandlish, Tom Henighan,
Tom B. Brown, Benjamin Chess, Rewon Child,
Scott Gray, Alec Radford, Jeffrey Wu, and Dario
Amodei. "Scaling Laws for Neural Language
Models." arXiv, January 23, 2020. https://arxiv.
org/abs/2001.08361.

Karnofsky, Holden. "Forecasting Transformative
AI, Part 1: What Kind of AI?" Cold Takes, August
10, 2021. https://www.cold-takes.com/
transformative-ai-timelines-part-1-of-4-what-
kind-of-ai/#impacts-of-pasta.

Karnofsky, Holden. "The Most Important Century."
Cold Takes, 2021. https://www.cold-takes.com/
most-important-century/.

Keles, Feyza Duman, Pruthuvi Mahesakya
Wijewardena, and Chinmay Hegde. "On the
Computational Complexity of Self-Attention."
Proceedings of Machine Learning Research 201
(2023): 1–23. https://proceedings.mlr.press/v201/
duman-keles23a/duman-keles23a.pdf.

Kenwood, A. G. "Railway Investment in Britain,
1825–1875." *Economica* 32, no. 127 (August 1965):
313–22. https://www.jstor.org/stable/2552228.

Khan, A.Q., and Munir Ahmad Khan. "Pakistani
Nuclear Program." Atomic Heritage Foundation,
August 23, 2018. https://ahf.nuclearmuseum.org/
ahf/history/pakistani-nuclear-program/.

Kirkpatrick, James, Razvan Pascanu, Neil
Rabinowitz, Joel Veness, Guillaume Desjardins,
Andrei A. Rusu, Kieran Milan, John Quan, Tiago
Ramalho, and Agnieszka Grabska-Barwinska.
"Overcoming Catastrophic Forgetting in Neural
Networks." arXiv, December 2, 2016. https://arxiv.
org/abs/1612.00796.

Knolle, Franziska. "Knowing What's Next: The Role
of the Cerebellum in Generating Predictions."
Max Planck Institute for Human Cognitive and
Brain Sciences, 2012. https://pure.mpg.de/rest/
items/item_1606219/component/file_1760316/
content.

Koch, Kristin, Judith McLean, Ronen Segev,
Michael A. Freed, Michael J. Berry II, Vijay
Balasubramanian, and Peter Sterling. "How
Much the Eye Tells the Brain." *Current
Biology* 16, no. 4 (2006): 1428–34. https://doi.
org/10.1016%2Fj.cub.2006.05.056.

Kocijan, Vid, Ernest Davis, Thomas Lukasiewicz,
Gary Marcus, and Leora Morgenstern. "The
Defeat of the Winograd Schema Challenge."
Artificial Intelligence 235 (2023): 103971. https://
doi.org/10.1016/j.artint.2023.103971.

Krakovna, Victoria, Jonathan Uesato, Vladimir
Mikulik, Matthew Rahtz, Tom Everitt, Ramana
Kumar, Zac Kenton, Jan Leike, and Shane
Legg. "Specification Gaming: The Flip Side
of AI Ingenuity." Google DeepMind, April 21,
2020. https://deepmind.google/discover/
blog/specification-gaming-the-flip-side-of-ai-
ingenuity/.

Kremer, Michael. "Population Growth and
Technological Change: One Million B.C. to 1990."
The Quarterly Journal of Economics 108, no. 3

(1993): 681–716. https://doi.org/10.2307/2118405.

Krizhevsky, Alex, Ilya Sutskever, and Geoffrey E. Hinton. "ImageNet Classification with Deep Convolutional Neural Networks." *Communications of the ACM* 60, no. 6 (2017): 84–90. https://doi.org/10.1145/3065386.

Kulveit, Jan, Clem von Stengel, and Roman Leventov. "Predictive Minds: LLMs as Atypical Active Inference Agents." arXiv, November 16, 2023. https://arxiv.org/abs/2311.10215.

Kurzweil, Ray. *The Age of Spiritual Machines.* Viking, 1999.

L

Lambert, Nathan. "The Q* Hypothesis: Tree-of-Thoughts Reasoning, Process Reward Models, and Supercharging Synthetic Data." Interconnects, November 24, 2023. https://www.interconnects.ai/p/q-star.

Lang, Kevin J., and Michael Witbrock. "Learning to Tell Two Spirals Apart." In *Proceedings of the 1998 Connectionist Models Summer School*, edited by D. Touretzky, G. Hinton, and T. Sejnowski. Morgan Kaufmann, 1988.

Lee, Jessica. "Did Richard Nixon Order Nuclear Strike on North Korea While Drunk?" Snopes, July 23, 2021. https://www.snopes.com/fact-check/north-korea-richard-nixon-nuclear/.

Leech, Gavin. "Mitchell's Open Problems." August 2, 2024. gleech.org/mitchell.

Leech, Gavin, Simson Garfinkel, Misha Yagudin, Alexander Brand, and Aleksandr Zhuravlev. "Ten Hard Problems in Artificial Intelligence We Must Get Right." arXiv, February 6, 2024. https://arxiv.org/abs/2402.04464.

Leech, Gavin, and Lynn. "Shallow Review of Live Agendas in Alignment & Safety." AI Alignment Forum, November 27, 2023. https://www.alignmentforum.org/posts/zaaGsFBeDTpCsYHef/shallow-review-of-live-agendas-in-alignment-and-safety.

Leech, Gavin, Juan J. Vazquez, Misha Yagudin, Niclas Kupper, and Laurence Aitchison. "Questionable Practices in Machine Learning." arXiv, July 17, 2024. https://arxiv.org/abs/2407.12220.

Legg, Shane, and Marcus Hutter. "Universal Intelligence: A Definition of Machine Intelligence." *Minds & Machines* 17 (2007): 391–444. https://doi.org/10.1007/s11023-007-9079-x.

Lewkowycz, Aitor, Anders Andreassen, David Dohan, Ethan Dyer, Henryk Michalewski, Vinay Ramasesh, Ambrose Slone, Cem Anil, Imanol Schlag, Theo Gutman-Solo, et al. "Solving Quantitative Reasoning Problems with Language Models." arXiv, June 29, 2022. https://arxiv.org/abs/2206.14858.

Li, Kenneth, Aspen K. Hopkins, David Bau, Fernanda Viégas, Hanspeter Pfister, and Martin Wattenberg. "Emergent World Representations: Exploring a Sequence Model Trained on a Synthetic Task." arXiv, October 24, 2022. https://arxiv.org/abs/2210.13382.

Lightman, Hunter, Vineet Kosaraju, Yura Burda, Harri Edwards, Bowen Baker, Teddy Lee, Jan

Leike, John Schulman, Ilya Sutskever, and Karl Cobbe. "Let's Verify Step by Step." arXiv, May 31, 2023. https://arxiv.org/abs/2305.20050.

Litan, Robert E. "The Telecommunications Crash: What to Do Now?" Brookings, December 1, 2002. https://www.brookings.edu/research/the-telecommunications-crash-what-to-do-now/.

Liu, Peter J., Mohammad Saleh, Etienne Pot, Ben Goodrich, Ryan Sepassi, Lukasz Kaiser, and Noam Shazeer. "Generating Wikipedia by Summarizing Long Sequences." arXiv, January 30, 2018. https://arxiv.org/abs/1801.10198.

López, María Martín. "Conceptual Mirrors: Reflecting on LLMs' Interpretations of Ideas." Berkeley D-Lab, April 23, 2024. https://dlab.berkeley.edu/news/conceptual-mirrors-reflecting-llms-interpretations-ideas.

Lovecraft, H. P. *Beyond the Wall of Sleep.* Arkham House, 1943.

Lupidi, Alisia, Carlos Gemmell, Nicola Cancedda, Jane Dwivedi-Yu, Jason Weston, Jakob Foerster, Roberta Raileanu, and Maria Lomeli. "Source2Synth: Synthetic Data Generation and Curation Grounded in Real Data Sources." arXiv, September 12, 2024. https://arxiv.org/abs/2409.08239.

Lyons, Austin. "Data Center Energy, National Security, Nuclear Power." Chipstrat, October 2, 2024. https://www.chipstrat.com/p/data-center-energy-needs-national.

Lyu, Qing, Shreya Havaldar, Adam Stein, Li Zhang, Delip Rao, Eric Wong, Marianna Apidianaki, and Chris Callison-Burch. "Faithful Chain-of-Thought Reasoning." *Proceedings of the 13th International Joint Conference on Natural Language Processing and the 3rd Conference of the Asia-Pacific Chapter of the Association for Computational Linguistics (Volume 1: Long Papers)* (2023): 305–29. https://aclanthology.org/2023.ijcnlp-main.20.pdf.

M

Madaan, Aman, Shuyan Zhou, Uri Alon, Yiming Yang, and Graham Neubig. "Language Models of Code Are Few-Shot Commonsense Learners." arXiv, October 13, 2022. https://arxiv.org/abs/2210.07128.

Mansour, Tarek. "Deep Neural Networks Are Lazy: On the Inductive Bias of Deep Learning." Master's thesis, Massachusetts Institute of Technology, 2018. https://dspace.mit.edu/bitstream/handle/1721.1/121680/1102057114-MIT.pdf.

Markowsky, George. "Applications of Information Theory: Physiology." Britannica, July 2, 2024. https://www.britannica.com/science/information-theory/Physiology.

Marr, David. *Vision: A Computational Approach.* Freeman & Co., 1982.

Maruf, Ramishah. "Google Fires Engineer Who Contended Its AI Technology Was Sentient." CNN, July 25, 2022. https://edition.cnn.com/2022/07/23/business/google-ai-engineer-fired-sentient/index.html.

McCullough, Brian. "A Revealing Look at the

Dot-Com Bubble of 2000—and How It Shapes Our Lives Today." TED, December 4, 2018. https://ideas.ted.com/an-eye-opening-look-at-the-dot-com-bubble-of-2000-and-how-it-shapes-our-lives-today/.

McGrath, Thomas, Matthew Rahtz, Janos Kramar, Vladimir Mikulik, and Shane Legg. "The Hydra Effect: Emergent Self-Repair in Language Model Computations." arXiv, July 28, 2023. https://arxiv.org/abs/2307.15771.

McIntosh, Stewart A., Fareeha Alam, Laura Adams, Ian S. Boon, Jonathan Callaghan, and Isabella Conti. "Global Funding for Cancer Research Between 2016 and 2020: A Content Analysis of Public and Philanthropic Investments." The Lancet Oncology 24, no. 6 (2023): 636–45. https://doi.org/10.1016/S1470-2045(23)00182-1.

Medawar, Peter. "Is the Scientific Paper a Fraud?" In The Strange Case of the Spotted Mice and Other Classic Essays on Science. Oxford University Press, 1996.

Merrill, William, and Ashish Sabharwal. "The Expressive Power of Transformers with Chain-of-Thought." arXiv, October 11, 2023. https://arxiv.org/abs/2310.07923.

Meta. "Introducing Meta Llama 3: The Most Capable Openly Available LLM to Date." April 18, 2024. https://ai.meta.com/blog/meta-llama-3/.

Metaculus. "Will There be a Complete 4 Year Interval in Which World Output Doubles, Before the First 1 Year Interval in Which World Output Doubles?" March 29, 2018. https://www.metaculus.com/questions/736/gwp-doubles-in-4-years-vs-1-year-by-2050/.

METR. "An Update on Our General Capability Evaluations." August 6, 2024. https://metr.org/blog/2024-08-06-update-on-evaluations/.

Metz, Cade. "In Two Moves, AlphaGo and Lee Sedol Redefined the Future." Wired, March 16, 2016. https://www.wired.com/2016/03/two-moves-alphago-lee-sedol-redefined-future/.

Microsoft and LinkedIn. "2024 Work Trend Index Annual Report." May 8, 2024. https://www.microsoft.com/en-us/worklab/work-trend-index/ai-at-work-is-here-now-comes-the-hard-part.

Miller, Evan. "Attention Is Off by One." July 24, 2023. https://www.evanmiller.org/attention-is-off-by-one.html.

Miller, Rich. "The Gigawatt Data Center Campus Is Coming." Data Center Frontier, April 29, 2024. https://www.datacenterfrontier.com/hyperscale/article/55021675/the-gigawatt-data-center-campus-is-coming.

Milmo, Dan. "ChatGPT Reaches 100 Million Users Two Months after Launch." The Guardian, February 2, 2023. https://www.theguardian.com/technology/2023/feb/02/chatgpt-100-million-users-open-ai-fastest-growing-app.

Milmo, Dan. "Impossible to Create AI Tools Like ChatGPT without Copyrighted Material." The Guardian, January 8, 2024. https://www.theguardian.com/technology/2024/jan/08/ai-tools-chatgpt-copyrighted-material-openai.

Milton, John. Areopagitica and Other Writings. Penguin, 2014.

Mirhoseini, Azalia, Anna Goldie, Mustafa Yazgan, Joe Wenjie Jiang, Ebrahim Songhori, Shen Wang, Young-Joon Lee, Eric Johnson, Omkar Pathak, Azade Nazi, et al. "A Graph Placement Methodology for Fast Chip Design." Nature 594 (2021): 207–12. https://doi.org/10.1038/s41586-021-03544-w.

Mitchell, Tom. "The Discipline of Machine Learning." Carnegie Mellon University, July 2006. https://www.cs.cmu.edu/~tom/pubs/MachineLearning.pdf.

Mnih, Volodymyr, Koray Kavukcuoglu, David Silver, Andrei A. Rusu, Joel Veness, Marc G. Bellemare, Alex Graves, Martin Riedmiller, Andreas K. Fidjeland, Georg Ostrovski, et al. "Human-Level Control through Deep Reinforcement Learning." Nature 518 (2015): 529–533. https://www.nature.com/articles/nature14236.

Mok, Aaron. "Amazon, Apple, and 12 Other Major Companies That Have Restricted Employees from Using ChatGPT." Business Insider, July 11, 2023. https://www.businessinsider.com/chatgpt-companies-issued-bans-restrictions-openai-ai-amazon-apple-2023-7.

Molloy, David. "The Great Graphics Card Shortage of 2020 (and 2021)." BBC News, January 24, 2021. https://www.bbc.co.uk/news/technology-55755820.

Moore, Gordon E. "Moore's Law at 40." In Understanding Moore's Law: Four Decades of Innovation, edited by David C. Brock. Chemical Heritage Foundation, 2006.

Morales, Jowi. "Elon Musk Powers New 'World's Fastest AI Data Center' with Gargantuan Portable Power Generators to Sidestep Electricity Supply Constraints." Tom's Hardware, July 24, 2024. https://www.tomshardware.com/tech-industry/artificial-intelligence/elon-musks-new-worlds-fastest-ai-data-center-is-powered-by-massive-portable-power-generators-to-sidestep-electricity-supply-constraints.

Moravec, Hans. "When Will Computer Hardware Match the Human Brain?" Journal of Evolution and Technology 1 (1998). https://jetpress.org/volume1/moravec.pdf.

Moulton, Ryan. "The Many Ways That Digital Minds Can Know." June 28, 2023. https://moultano.wordpress.com/2023/06/28/the-many-ways-that-digital-minds-can-know/.

Mu, Tong, Alec Helyar, Johannes Heidecke, Joshua Achiam, Andrea Vallone, Ian Kivlichan, Molly Lin, Alex Beutel, John Schulman, and Lilian Weng. "Rule-Based Rewards for Language Model Safety." OpenAI, July 24, 2024. https://cdn.openai.com/rule-based-rewards-for-language-model-safety.pdf.

N

Nadella, Satya. "Microsoft Ignite Opening." Microsoft, November 15, 2023. https://news.microsoft.com/wp-content/uploads/prod/2023/11/Microsoft-Ignite-Opening.pdf.

Nakano, Kaoru. "Learning Process in a Model of

Associative Memory." In *Pattern Recognition and Machine Learning*, edited by K.S. Fu. Springer, 1971.

Nakano, Reiichiro, Jacob Hilton, Suchir Balaji, Jeff Wu, Long Ouyang, Christina Kim, Christopher Hesse, Shantanu Jain, Vineet Kosaraju, William Saunders, et al. "WebGPT: Browser-Assisted Question-Answering with Human Feedback." arXiv, December 17, 2021. https://arxiv.org/abs/2112.09332.

Nanda, Neel, Lawrence Chan, Tom Lieberum, Jess Smith, and Jacob Steinhardt. "Progress Measures for Grokking via Mechanistic Interpretability." arXiv, January 12, 2023. https://arxiv.org/abs/2301.05217.

Nanda, Neel, Senthooran Rajamanoharan, János Kramár, and Rohin Shah. "Fact Finding: Attempting to Reverse-Engineer Factual Recall on the Neuron Level." AI Alignment Forum, December 23, 2023. https://www.alignmentforum.org/posts/iGuwZTHWb6DFY3sKB/fact-finding-attempting-to-reverse-engineer-factual-recall.

NASA. "FY 2023 Budget Request." February 2023. https://www.nasa.gov/wp-content/uploads/2023/02/fy23_nasa_budget_request_summary.pdf?emrc=b2750f.

National Institutes of Health. "Budget." Last updated July 30, 2024. https://www.nih.gov/about-nih/what-we-do/budget.

National Science Foundation. "Budget, Performance, and Financial Reporting." 2024. https://new.nsf.gov/about/budget.

Nedelkoska, Ljubica, and Glenda Quintini. "Automation, Skills Use, and Training." OECD Social, Employment and Migration Working Papers, No. 202. OECD Publishing, 2018. https://doi.org/10.1787/2e2f4eea-en.

Newell, Allen, John C. Shaw, and Herbert A. Simon. "Report on a General Problem-Solving Program." The Rand Corporation, December 30, 1958. https://exhibits.stanford.edu/feigenbaum/catalog/sy501xd1313.

Nezhurina, Marianna, Lucia Cipolina-Kun, Mehdi Cherti, and Jenia Jitsev. "Alice in Wonderland: Simple Tasks Showing Complete Reasoning Breakdown in State-of-the-Art Large Language Models." arXiv, June 4, 2024. https://arxiv.org/abs/2406.02061.

Ng, Andrew. "Issue 242." The Batch, March 27, 2024. https://www.deeplearning.ai/the-batch/issue-242/.

Niplav. "Transfer Learning in Humans." LessWrong, April 21, 2024. https://www.lesswrong.com/posts/QTTCRytvyFteJgPwg/transfer-learning-in-humans.

Nori, Harsha, Yin Tat Lee, Sheng Zhang, Dean Carignan, Richard Edgar, Nicolo Fusi, Nicholas King, Jonathan Larson, Yuanzhi Li, Weishung Liu, et al. "Can Generalist Foundation Models Outcompete Special-Purpose Tuning? Case Study in Medicine." arXiv, November 28, 2023. https://arxiv.org/abs/2311.16452.

Nostalgebraist. "GPT-3: A Disappointing Paper." LessWrong, May 29, 2020. https://www.lesswrong.com/posts/ZHrpjDc3CepSeeBuE/gpt-3-a-disappointing-paper.

Nostalgebraist. "The Scaling 'Inconsistency': OpenAI's New Insight." Trees Are Harlequins, Words Are Harlequins, 2020. https://nostalgebraist.tumblr.com/post/634109483113676800/the-scaling-inconsistency-openais-new-insight.

Notopoulos, Katie. "A Chevy Dealership Added an AI Chatbot to Its Site. Then All Hell Broke Loose." Business Insider, December 18, 2023. https://www.businessinsider.com/car-dealership-chevrolet-chatbot-chatgpt-pranks-chevy-2023-12.

Nozick, Robert. *Anarchy, State, and Utopia*. Basic Books, 1974.

Nuclear Newswire. "Amazon Buys Nuclear-Powered Data Center from Talen." March 7, 2024. https://www.ans.org/news/article-5842/amazon-buys-nuclearpowered-data-center-from-talen/.

O

Olah, Chris, Nick Cammarata, Ludwig Schubert, Gabriel Goh, Michael Petrov, and Shan Carter. "Zoom In: An Introduction to Circuits." Distill, March 10, 2020. https://doi.org/10.23915/distill.00024.001.

Olah, Christopher, Alexander Mordvintsev, and Ludwig Schubert. "Feature Visualization." Distill, November 7, 2017. https://distill.pub/2017/feature-visualization.

Olds, James, and Peter Milner. "Positive Reinforcement Produced by Electrical Stimulation of Septal Area and Other Regions of Rat Brain." *Journal of Comparative and Physiological Psychology* 47, no. 6 (1954): 4191–27. https://doi.org/10.1037/h0058775.

Olshausen, Bruno A., and David J. Field. "Sparse Coding with an Overcomplete Basis Set: A Strategy Employed by V1?" *Vision Research* 37, no. 23 (1997): 3311–25. https://doi.org/10.1016/S0042-6989(97)00169-7.

Olsson, Catherine, Nelson Elhage, Neel Nanda, Nicholas Joseph, Nova DasSarma, Tom Henighan, Ben Mann, Amanda Askell, Yuntao Bai, Anna Chen, et al. "In-Context Learning and Induction Heads." Transformer Circuits Thread, March 8, 2022. https://transformer-circuits.pub/2022/in-context-learning-and-induction-heads/index.html.

Omohundro, Stephen M. "The Basic AI Drives." *Proceedings of the 2008 Conference on Artificial General Intelligence 2008* (2008): 483–92. https://dl.acm.org/doi/10.5555/1566174.1566226.

OpenAI. "A Landmark Multi-Year Partnership with News Corp." May 22, 2024. https://openai.com/index/news-corp-and-openai-sign-landmark-multi-year-global-partnership/.

OpenAI. "GPT-4 Technical Report." arXiv, March 15, 2023. https://arxiv.org/abs/2303.08774.

OpenAI. "OpenAI and Reddit Partnership." May 16, 2024. https://openai.com/index/openai-and-reddit-partnership/.

OpenAI. "OpenAI o1 System Card." September 12, 2024. https://assets.ctfassets.net/kftzwdyauwt9/67qJD51Aur3elc96iOfeOP/71551c3d223cd97e591aa89567306912/o1_system_card.pdf.

OpenAI. "Preparedness Framework (Beta)." December 18, 2023. https://cdn.openai.com/openai-preparedness-framework-beta.pdf.

Our World in Data. "Global GDP Over the Long Run." Last updated May 16, 2024. https://ourworldindata.org/grapher/global-gdp-over-the-long-run.

Our World in Data. "Moore's Law: The Number of Transistors Per Microprocessor." 2022. https://ourworldindata.org/grapher/transistors-per-microprocessor.

Owen, David. "Extrapolating Performance in Language Modeling Benchmarks." Epoch AI, June 12, 2023. https://epochai.org/files/llm-benchmark-extrapolation.pdf.

P

Patel, Dylan, and Daniel Nishball. "Google Gemini Eats the World—Gemini Smashes GPT-4 by 5x, the GPU-Poors." SemiAnalysis, August 28, 2023. https://www.semianalysis.com/p/google-gemini-eats-the-world-gemini.

Patel, Dylan, Daniel Nishball, and Jeremie Eliahou Ontiveros. "AI Datacenter Energy Dilemma—Race for AI Datacenter Space." SemiAnalysis, March 13, 2024. https://www.semianalysis.com/p/ai-datacenter-energy-dilemma-race.

Patronus AI. "Introducing CopyrightCatcher, the First Copyright Detection API for LLMs." March 6, 2024. https://www.patronus.ai/blog/introducing-copyright-catcher.

Perez, Jorge, Javier Marinkovic, and Pablo Barceló. "On the Turing Completeness of Modern Neural Network Architectures." arXiv, January 10, 2019. https://arxiv.org/abs/1901.03429.

Perez, Sarah. "AI App Character.ai Is Catching Up to ChatGPT in the US." TechCrunch, September 11, 2023. https://techcrunch.com/2023/09/11/ai-app-character-ai-is-catching-up-to-chatgpt-in-the-u-s/.

Perrigo, Billy. "Bing's AI Is Threatening Users. That's No Laughing Matter." Time, February 17, 2023. https://time.com/6256529/bing-openai-chatgpt-danger-alignment/.

Perry, Mark J. "Looking Back at the Remarkable History of the Nobel Prize from 1901–2020 Using Maps, Charts, and Tables." AEI, October 12, 2020. https://www.aei.org/carpe-diem/looking-back-at-the-remarkable-history-of-the-nobel-prize-from-1901-2020-using-maps-charts-and-tables/.

Pezeshkpour, Pouya, and Estevam Hruschka. "Large Language Models Sensitivity to the Order of Options in Multiple-Choice Questions." arXiv, August 22, 2023. https://arxiv.org/abs/2308.11483.

Pfau, Jacob. "Will the ARC-AGI Grand Prize Be Claimed by the End of 2025?" Manifold. https://manifold.markets/JacobPfau/will-the-arcagi-grand-prize-be-clai-srb6t2awj1.

Piantadosi, Steven T., Dyana C.Y. Muller, Joshua S.

Rule, Karthikeya Kaushik, Mark Gorenstein, Elena R. Leib, and Emily Sanford. "Why Concepts Are (Probably) Vectors." Trends in Cognitive Sciences (2024). https://doi.org/10.1016/j.tics.2024.06.011.

Pichai, Sundar. "Q3 Earnings Call: CEO's Remarks." The Keyword, October 29, 2024. https://blog.google/inside-google/message-ceo/alphabet-earnings-q3-2024.

Pinker, Steven, and Alan Prince. "On Language and Connectionism: Analysis of a Parallel Distributed Processing Model of Language Acquisition." Cognition 28, no. 1–2 (March 1988): 73–193. https://doi.org/10.1016/0010-0277(88)90032-7.

Plappert, Matthias, Durk Kingma, Max Chen, Cage Zhong, and Penny Deng. "Thoughts on Llama 3." Factorial Funds, April 24, 2024. https://www.factorialfunds.com/blog/thoughts-on-llama-3.

Plato, Republic. Translated by Robin Waterfield. Oxford University Press, 1993.

Pope, Reiner, Sholto Douglas, Aakanksha Chowdhery, Jacob Devlin, James Bradbury, Jonathan Heek, Kefan Xiao, Shivani Agrawal, and Jeff Dean. "Efficiently Scaling Transformer Inference." Proceedings of the 6th MLSys Conference, 2023. https://proceedings.mlsys.org/paper_files/paper/2023/hash/c4be71ab8d24cdfb45e3d06dbfca2780-Abstract-mlsys2023.html.

Power, Alethea, Yuri Burda, Harri Edwards, Igor Babuschkin, and Vedant Misra. "Grokking: Generalization Beyond Overfitting on Small Algorithmic Dataset." arXiv, January 6, 2022. https://arxiv.org/abs/2201.02177.

Protzko, John, and Roberto Colom. "Testing the Structure of Human Cognitive Ability Using Evidence Obtained from the Impact of Brain Lesions over Abilities." Intelligence 89 (November–December 2021): 101581. https://doi.org/10.1016/j.intell.2021.101581.

Q

Qi, Xiangyu, Ashwinee Panda, Kaifeng Lyu, Xiao Ma, Subhrajit Roy, Ahmad Beirami, Prateek Mittal, and Peter Henderson. "Safety Alignment Should Be Made More Than Just a Few Tokens Deep." arXiv, June 10, 2024. https://arxiv.org/abs/2406.05946.

R

Radford, Alec, Jong Wook Kim, Chris Hallacy, Aditya Ramesh, Gabriel Goh, Sandhini Agarwal, Girish Sastry, Amanda Askell, Pamela Mishkin, Jack Clark, et al. "Learning Transferable Visual Models from Natural Language Supervision." arXiv, February 26, 2021. https://arxiv.org/abs/2103.00020.

Rafailov, Rafael, Archit Sharma, Eric Mitchell, Stefano Ermon, Christopher D. Manning, and Chelsea Finn. "Direct Preference Optimization: Your Language Model Is Secretly a Reward Model." arXiv, May 29, 2023. https://arxiv.org/abs/2305.18290.

Raven, John C. "Standardization of Progressive Matrices, 1938." *Psychology and Psychotherapy* 19, no. 1 (1941): 137–50. https://doi.org/10.1111/j.2044-8341.1941.tb00316.x.

Raymond, Eric S. "Some AI Koans." catb.org. http://www.catb.org/esr/jargon/html/koans.html.

Recht, Benjamin, Rebecca Roelofs, Ludwig Schmidt, and Vaishaal Shankar. "Do ImageNet Classifiers Generalize to ImageNet?" arXiv, February 13, 2019. https://arxiv.org/abs/1902.10811/.

Ringel Morris, Meredith, Jascha Sohl-Dickstein, Noah Fiedel, Tris Warkentin, Allan Dafoe, Aleksandra Faust, Clement Farabet, and Shane Legg. "Levels of AGI for Operationalizing Progress on the Path to AGI." *Proceedings of ICML 2024.* https://arxiv.org/abs/2311.02462.

Robinson, John A. "A Machine-Oriented Logic Based on the Resolution Principle." *Journal of the ACM* 12, no. 1 (1965): 23–41. https://doi.org/10.1145/321250.321253.

Roger, Fabien, and Ryan Greenblatt. "Preventing Language Models from Hiding Their Reasoning." arXiv, October 27, 2023. https://arxiv.org/abs/2310.18512.

Rogers, Tiffany D., Eric McKimm, Price E. Dickson, Dan Goldowitz, Charles D. Blaha, and Guy Mittleman. "Is Autism a Disease of the Cerebellum? An Integration of Clinical and Pre-Clinical Research." *Frontiers in Systems Neuroscience* 7 (2013): 15. https://doi.org/10.3389%2Ffnsys.2013.00015.

Roose, Kevin. "Why a Conversation with Bing's Chatbot Left Me Deeply Unsettled." *The New York Times*, February 16, 2023. https://www.nytimes.com/2023/02/16/technology/bing-chatbot-microsoft-chatgpt.html.

Roose, Kevin. "Why an Octopus-Like Creature Has Come to Symbolize the State of AI." *The New York Times*, May 30, 2023. https://www.nytimes.com/2023/05/30/technology/shoggoth-meme-ai.html.

Roy, Rajarshi, Jonathan Raiman, Neel Kant, Ilyas Elkin, Robert Kirby, Michael Siu, Stuart Oberman, Saad Godil, and Bryan Catanzaro. "PrefixRL: Optimization of Parallel Prefix Circuits Using Deep Reinforcement Learning." arXiv, May 14, 2022. https://arxiv.org/abs/2205.07000.

"Roy Amara." In *Oxford Essential Quotations*, edited by Susan Ratcliffe. Oxford University Press, 2016. https://www.oxfordreference.com/display/10.1093/acref/9780191826719.001.0001/q-oro-ed4-00018679.

Royal College of Radiologists. "2021 Clinical Radiology Census Report." 2021. https://www.rcr.ac.uk/media/30dhjeh2/clinical_radiology_census_report_2021.pdf.

Royal College of Radiologists. "2022 Clinical Radiology Workforce Census." 2022. https://www.rcr.ac.uk/media/qs0jnfmv/rcr-census_clinical-radiology-workforce-census_2022.pdf.

Ruoss, Anian, Grégoire Delétang, Sourabh Medapati, Jordi Grau-Moya, Li Kevin Wenliang, Elliot Catt, John Reid, and Tim Genewein. "Grandmaster-Level Chess without Search." arXiv, February 7, 2024. https://arxiv.org/abs/2402.04494.

S

Sang-Deok, Lee, and Lee Deok-ju. "오픈AI, 물리학 난제 도전 범용과학 인공지능 만든다." Maeil Kyungjae, March 3, 2024. https://n.news.naver.com/article/009/0005266676.

Sanzo, Kameron. "What Came First: Thermodynamics or the Steam Engine?" Medium, September 14, 2020. https://medium.com/@kameron.sanzo/what-came-first-thermodynamics-or-the-steam-engine-c0cca0996b0b.

Schaeffer, Rylan, Brando Miranda, and Sanmi Koyejo. "Are Emergent Abilities of Large Language Models a Mirage?" arXiv, April 28, 2023. https://arxiv.org/abs/2304.15004.

Schaeffer, Rylan, Hailey Schoelkopf, Brando Miranda, Gabriel Mukobi, Varun Madan, Adam Ibrahim, Herbie Bradley, Stella Biderman, and Sanmi Koyejo. "Why Has Predicting Downstream Capabilities of Frontier AI Models with Scale Remained Elusive?" ICML 2024 Next Generation of AI Safety Workshop, Vienna, Austria, July 26, 2024. https://openreview.net/forum?id=OuD8PFGbfN.

Schnitzer, Monika, Martin Watzinger, and Markus Nagler. "Fostering the Diffusion of General Purpose Technologies: Evidence from the Transistor." CEPR VoxEU, February 8, 2021. https://cepr.org/voxeu/columns/fostering-diffusion-general-purpose-technologies-evidence-transistor.

Schreiner, Maximilian. "GPT-4 Architecture, Datasets, Costs and More Leaked." The Decoder, July 11, 2023. https://the-decoder.com/gpt-4-architecture-datasets-costs-and-more-leaked/.

Schroeder, Stan. "OpenAI Chief Scientist Ilya Sutskever Is Leaving. But What Did He See?" Mashable, May 15, 2024. https://mashable.com/article/openair-ilya-slutskever-leaves-chief-scientist.

Schubert, Stefan, Lucius Caviola, and Nadira S. Faber. "The Psychology of Existential Risk: Moral Judgments about Human Extinction." *Scientific Reports* 9 (2019). https://www.nature.com/articles/s41598-019-50145-9.

Schulman, John. "Optimizing Expectations: From Deep Reinforcement Learning to Stochastic Computation Graphs." University of California at Berkeley Technical Report No. UCB/EECS-2016-217, December 16, 2016. https://www2.eecs.berkeley.edu/Pubs/TechRpts/2016/EECS-2016-217.pdf.

Schulman, John, Filip Wolski, Prafulla Dhariwal, Alec Radford, and Oleg Klimov. "Proximal Policy Optimization Algorithms." arXiv, July 20, 2017. https://arxiv.org/abs/1707.06347.

Semrush. "Most Visited Websites in the World." Last updated July 2024. https://www.semrush.com/website/top/.

Sevilla, Jaime, Tamay Besiroglu, Ben Cottier, Josh You, Edu Roldán, Pablo Villalobos, and Ege Erdil. "Can AI Scaling Continue through 2030?" Epoch AI, 2024. https://epochai.org/blog/can-ai-scaling-continue-through-2030#synthetic-data.

Sevilla, Jaime, Lennart Heim, Anson Ho, Tamay

Besiroglu, Marius Hobbhahn, and Pablo Villalobos. "Compute Trends Across Three Eras of Machine Learning." Epoch AI, February 16, 2022. https://epochai.org/blog/compute-trends.

Sevilla, Jaime, and Edu Roldán. "Training Compute of Frontier AI Models Grows by 4–5x per Year." Epoch AI, May 28, 2024. https://epochai.org/blog/training-compute-of-frontier-ai-models-grows-by-4-5x-per-year.

Shah, Rusheb, Quentin Feuillade-Montixi, Soroush Pour, Arush Tagade, Stephen Casper, and Javier Rando. "Scalable and Transferable Black-Box Jailbreaks for Language Models via Persona Modulation." arXiv, November 6, 2023. https://arxiv.org/abs/2311.03348.

Shalizi, Cosma. "So You Think You Have a Power Law—Well Isn't That Special?" Three-Toed Sloth, June 15, 2007. http://bactra.org/weblog/491.html.

Sharma, Utkarsh, and Jared Kaplan. "Scaling Laws from the Data Manifold Dimension." *Journal of Machine Learning Research* 23, no. 9 (2022) 1–34. https://jmlr.csail.mit.edu/papers/volume23/20-1111/20-1111.pdf.

Shlegeris, Buck, Fabien Roger, Lawrence Chan, and Euan McLean. "Language Models Are Better Than Humans at Next-Token Prediction." arXiv, December 21, 2022. https://arxiv.org/abs/2212.11281.

Shulman, Carl, and Nick Bostrom. "How Hard Is Artificial Intelligence? Evolutionary Arguments and Selection Effects." *Journal of Consciousness Studies* 19, no. 7–8 (2012): 103–30. https://nickbostrom.com/aievolution.pdf.

Shumailov, Ilia, Zakhar Shumaylov, Yiren Zhao, Nicolas Papernot, Ross Anderson, and Yarin Gal. "AI Models Collapse When Trained on Recursively Generated Data." *Nature* 631 (2024): 755–59. https://doi.org/10.1038/s41586-024-07566-y.

Silver, David, Aja Huang, Chris J. Maddison, Arthur Guez, Laurent Sifre, George van den Driessche, Julian Schrittwieser, Ioannis Antonoglou, Veda Panneershelvam, Marc Lanctot, et al."Mastering the Game of Go with Deep Neural Networks and Tree Search." *Nature* 529 (2016): 484–89. https://doi.org/10.1038/nature16961.

Silver, David, Thomas Hubert, Julian Schrittwieser, Ioannis Antonoglou, Matthew Lai, Arthur Guez, Marc Lanctot, Laurent Sifre, Dharshan Kumaran, Thore Graepel, et al. "Mastering Chess and Shogi by Self-Play with a General Reinforcement Learning Algorithm." arXiv, December 5, 2017. https://arxiv.org/abs/1712.01815.

Silver, David, Julian Schrittwieser, Karen Simonyan, Ioannis Antonoglou, Aja Huang, Arthur Guez, Thomas Hubert, Lucas Baker, Matthew Lai, Adrian Bolton, et al. "Mastering the Game of Go without Human Knowledge." *Nature* 550 (2017): 354–59. https://doi.org/10.1038/nature24270.

"Size of Wikipedia." Wikipedia. Last updated November 20, 2024. https://en.wikipedia.org/wiki/Wikipedia:Size_of_Wikipedia.

Skalse, Joar, Nikolaus H. R. Howe, Dmitrii Krasheninnikov, and David Krueger. "Defining and Characterizing Reward Hacking." arXiv, September 27, 2022. https://arxiv.org/abs/2209.13085.

Snell, Charlie, Jaehoon Lee, Kelvin Xu, and Aviral Kumar. "Scaling LLM Test-Time Compute Optimally Can Be More Effective Than Scaling Model Parameters." arXiv, August 6, 2024. https://arxiv.org/abs/2408.03314.

Sperry, Douglas E., Linda L. Sperry, and Peggy J. Miller. "Reexamining the Verbal Environments of Children from Different Socioeconomic Backgrounds." *Child Development* 90, no. 4 (2019): 1303–18. https://doi.org/10.1111/cdev.13072.

Srivastava, Saurabh, Annarose M B, Anto P V, Shashank Menon, Ajay Sukumar, Adwaith Samod T, Alan Philipose, Stevin Prince, and Sooraj Thomas. "Functional Benchmarks for Robust Evaluation of Reasoning Performance, and the Reasoning Gap." arXiv, February 29, 2024. https://arxiv.org/abs/2402.19450.

Stanovich, Keith E., and Richard F. West. "Individual Difference in Reasoning: Implications for the Rationality Debate?" *Behavioral and Brain Sciences* 23, no. 5 (2001): 645–65. https://doi.org/10.1017/s0140525x00003435.

Stiennon, Nisan, Long Ouyang, Jeff Wu, Daniel M. Ziegler, Ryan Lowe, Chelsea Voss, Alec Radford, Dario Amodei, and Paul Christiano. "Learning to Summarize with Human Feedback." Advances in Neural Information Processing Systems 33 (NeurIPS 2020). https://proceedings.neurips.cc/paper_files/paper/2020/file/1f89885d556929e98d3ef9b86448f951-Paper.pdf.

Stockfish. "Stockfish 16.1." February 24, 2024. https://stockfishchess.org/blog/2024/stockfish-16-1/.

Strausfeld, Nicholas J., Lars Hansen, Yongsheng Li, Robert S. Gomez, and Kei Ito. "Evolution, Discovery, and Interpretations of Arthropod Mushroom Bodies." *Learning & Memory* 5, no. 1 (1998): 11–37. https://www.ncbi.nlm.nih.gov/pmc/articles/PMC311242/.

Stribling, Daniel, Yuxing Xia, Maha K. Amer, Kiley S. Graim, Connie J. Mulligan, and Rolf Renne. "The Model Student: GPT-4 Performance on Graduate Biomedical Science Exams." *Scientific Reports* 14 (2024). https://doi.org/10.1038/s41598-024-55568-7.

Stromberg, Joseph. "The Neuroscientist Who Discovered He Was a Psychopath." *Smithsonian Magazine*, November 22, 2013. https://www.smithsonianmag.com/science-nature/the-neuroscientist-who-discovered-he-was-a-psychopath-180947814/.

Sun, Xingwu, Yanfeng Chen, Yiqing Huang, Ruobing Xie, Jiaqi Zhu, Kai Zhang, Shuaipeng Li, Zhen Yang, Jonny Han, Xiaobo Shu, et al. "Hunyuan-Large: An Open-Source MoE Model with 52 Billion Activated Parameters by Tencent." arXiv, November 4, 2024. https://arxiv.org/abs/2411.02265.

Sutskever, Ilya, Oriol Vinyals, and Quoc V. Le. "Sequence to Sequence Learning with Neural Networks." arXiv, September 10, 2014. https://arxiv.org/abs/1409.3215.

Sutton, Richard S. "The Bitter Lesson." March 13, 2019. http://www.incompleteideas.net/IncIdeas/BitterLesson.html.

Synced. "Yann LeCun Cake Analogy 2.0." February

22, 2019. https://syncedreview.com/2019/02/22/yann-lecun-cake-analogy-2-0/.

T

Tabachnyk, Maxim, and Stoyan Nikolov. "ML-Enhanced Code Completion Improves Developer Productivity." Google Research, July 26, 2022. https://research.google/blog/ml-enhanced-code-completion-improves-developer-productivity/.

Templeton, Adly, Tom Conerly, Jonathan Marcus, Jack Lindsey, Trenton Bricken, Brian Chen, Adam Pearce, Craig Citro, Emmanuel Ameisen, Andy Jones, et al. "Scaling Monosemanticity: Extracting Interpretable Features from Claude 3 Sonnet." Transformer Circuits Thread, May 21, 2024. https://transformer-circuits.pub/2024/scaling-monosemanticity/index.html.

Tigges, Curt, Michael Hanna, Qinan Yu, and Stella Biderman. "LLM Circuit Analyses Are Consistent Across Training and Scale." arXiv, July 15, 2024. https://arxiv.org/abs/2407.10827.

Topol, Eric. "Geoffrey Hinton: Large Language Models in Medicine. They Understand and Have Empathy." Ground Truths, December 8, 2023. https://erictopol.substack.com/p/geoffrey-hinton-large-language-models.

Trammell, Philip, and Anton Korinek. "Economic Growth under Transformative AI." GPI Working Paper No. 8-2020. https://globalprioritiesinstitute.org/wp-content/uploads/Philip-Trammell-and-Anton-Korinek_economic-growth-under-transformative-ai.pdf.

Turing, Alan. "Computing Machinery and Intelligence." *Mind* LIX, no. 236 (October 1950): 433–60.

Turing, Alan. "Intelligent Machinery, A Heretical Theory (c. 1951)." In *The Essential Turing,* edited by B. J. Copeland. Oxford Academic, 2004. https://doi.org/10.1093/oso/9780198250791.003.0018.

U

US Energy Information Administration. "How Much Natural Gas Does the United States Have, and How Long Will It Last?" Last updated April 29, 2024. https://www.eia.gov/tools/faqs/faq.php?id=58.

US Energy Information Administration. "New Natural Gas-Fired Capacity Additions Expected to Total 8.6 Gigawatts in 2023." October 16, 2023. https://www.eia.gov/todayinenergy/detail.php?id=60663.

US Energy Information Administration. "What Is US Electricity Generation by Energy Source?" Last updated February 29, 2024. https://www.eia.gov/tools/faqs/faq.php?id=427.

US Senate Subcommittee on the Judiciary. "Oversight of AI: Rules for Artificial Intelligence." May 16, 2023. Serial No. J–118–16.https://www.govinfo.gov/content/pkg/CHRG-118shrg52706/pdf/CHRG-118shrg52706.pdf.

V

Vafa, Keyon, Justin Y. Chen, Jon Kleinberg, Sendhil Mullainathan, and Ashesh Rambachan. "Evaluating the World Model Implicit in a Generative Model." arXiv, June 6, 2024. https://arxiv.org/abs/2406.03689.

Valmeekam, Karthik, Kaya Stechly, Atharva Gundawar, and Subbarao Kambhampati. "Planning in Strawberry Fields: Evaluating and Improving the Planning and Scheduling Capabilities of LRM o1." arXiv, October 3, 2024. https://arxiv.org/abs/2410.02162.

van der Weij, Teun, Felix Hofstätter, Oliver Jaffe, and Francis Rhys Ward. "AI Sandbagging: Language Models Can Strategically Underperform on Evaluations." arXiv, June 11, 2024. https://arxiv.org/html/2406.07358v3.

van Gilst, Koen. "Analyzing GPT-4 Tokens." May 2, 2024. https://koenvangilst.nl/blog/analyzing-gpt-4-tokens.

van Schaik, Tempest A., and Brittany Pugh. "A Field Guide to Automatic Evaluation of LLM-Generated Summaries." SIGIR '24: Proceedings of the 47th International ACM SIGIR Conference on Research and Development in Information Retrieval (July 2024). https://dl.acm.org/doi/abs/10.1145/3626772.3661346.

Vaswani, Ashish, Noam Shazeer, Niki Parmar, Jakob Uszkoreit, Llion Jones, Aidan N. Gomez, Łukasz Kaiser, and Illia Polosukhin. "Attention Is All You Need." arXiv, June 12, 2017. https://arxiv.org/abs/1706.03762.

Verma, Pranshu. "They Fell in Love with AI Bots. A Software Update Broke Their Hearts." *The Washington Post*, March 30, 2023. https://www.washingtonpost.com/technology/2023/03/30/replika-ai-chatbot-update/.

Vicuna Team. "Vicuna: An Open-Source Chatbot Impressing GPT-4 with 90%* ChatGPT Quality." LMSYS Org, March 30, 2023. https://lmsys.org/blog/2023-03-30-vicuna/.

Villalobos, Pablo, Anson Ho, Jaime Sevilla, Tamay Besiroglu, Lennart Heim, and Marius Hobbhahn. "Will We Run Out of Data? Limits of LLM Scaling Based on Human-Generated Data." arXiv, October 26, 2022. https://arxiv.org/abs/2211.04325.

Villalobos, Pablo, Jaime Sevilla, Lennart Heim, Tamay Besiroglu, Marius Hobbhahn, and Anson Ho. "Will We Run Out of ML Data? Evidence from Projecting Dataset Size Trends." Epoch AI, November 10, 2022. https://epochai.org/blog/will-we-run-out-of-ml-data-evidence-from-projecting-dataset.

Vincent, James. "ChatGPT Proves AI Is Finally Mainstream—And Things Are Only Going to Get Weirder." *The Verge*, December 8, 2022. https://www.theverge.com/2022/12/8/23499728/ai-capability-accessibility-chatgpt-stable-diffusion-commercialization.

Vinge, Vernor. "The Coming Technological Singularity: How to Survive in the Post-Human Era." NASA Lewis Research Center, Vision 21: Interdisciplinary Science and Engineering in the Era of Cyberspace, December 1, 1993. https://ntrs.nasa.gov/citations/19940022856.

W

Walsh, Bryan. "60 Years Ago Today, This Man Stopped the Cuban Missile Crisis from Going Nuclear." *Vox*, October 27, 2022. https://www.vox.com/future-perfect/2022/10/27/23426482/cuban-missile-crisis-basilica-arkhipov-nuclear-war.

Walton Family Foundation. "The Value of AI in Today's Classrooms." June 11, 2024. https://www.waltonfamilyfoundation.org/learning/the-value-of-ai-in-todays-classrooms.

Wang, Shuohang, Yang Liu, Yichong Xu, Chenguang Zhu, and Michael Zeng. "Want to Reduce Labeling Cost? GPT-3 Can Help." arXiv, August 30, 2021. https://arxiv.org/abs/2108.13487.

Wang, Zhengdong. "Why Transformative AI Is Really, Really Hard to Achieve." June 27, 2023. https://zhengdongwang.com/2023/06/27/why-transformative-ai-is-really-really-hard-to-achieve.html.

Warden, Pete. "The Death of Feature Engineering Is Greatly Exaggerated." December 11, 2021. https://petewarden.com/2021/12/11/the-death-of-feature-engineering-is-greatly-exaggerated/.

Watson, Stuart K., Judith M. Burkart, Steven J. Schapiro, Susan P. Lambeth, Jutta L. Mueller, and Simon W. Townsend. "Nonadjacent Dependency Processing in Monkeys, Apes, and Humans." *Science Advances* 6, no. 43 (2020). https://doi.org/10.1126/sciadv.abb0725.

Wei, Alexander, Nika Haghtalab, and Jacob Steinhardt. "Jailbroken: How Does LLM Safety Training Fail?" In *Advances in Neural Information Processing Systems 36 (NeurIPS 2023)*, edited by A. Oh, T. Naumann, A. Globerson, K. Saenko, M. Hardt, and S. Levine. 2023. https://proceedings.neurips.cc/paper_files/paper/2023/hash/fd6613131889a4b656206c50a8bd7790-Abstract-Conference.html.

Wei, Jason, Xuezhi Wang, Dale Schuurmans, Maarten Bosma, Brian Ichter, Fei Xia, Ed H. Chi, Quoc V. Le, and Denny Zhou. "Chain-of-Thought Prompting Elicits Reasoning in Large Language Models." arXiv, January 28, 2022. https://arxiv.org/abs/2201.11903.

Weijers, Dan. "Nozick's Experience Machine Is Dead, Long Live the Experience Machine!" *Philosophical Psychology* 27, no. 4 (February 2014): 513–35. https://www.tandfonline.com/doi/full/10.1080/09515089.2012.757889.

Wentworth, John S. "Godzilla Strategies." LessWrong, June 11, 2022. https://www.lesswrong.com/posts/DwqgLXn5qYC7GqExF/godzilla-strategies.

Werbos, Paul J. "Applications of Advances in Nonlinear Sensitivity Analysis." *System Modeling and Optimization: Proceedings of the 10th IFIP Conference New York City* (1981): 762–70. https://link.springer.com/chapter/10.1007/BFb0006203.

The White House. "Memorandum on Advancing the United States' Leadership in Artificial Intelligence; Harnessing Artificial Intelligence to Fulfill National Security Objectives; and Fostering the Safety, Security, and Trustworthiness of Artificial Intelligence." October 24, 2024. https://www.whitehouse.gov/briefing-room/presidential-actions/2024/10/24/memorandum-on-advancing-the-united-states-leadership-in-artificial-intelligence-harnessing-artificial-intelligence-to-fulfill-national-security-objectives-and-fostering-the-safety-security/.

Willison, Simon. "Notes on OpenAI's New o1 Chain-of-Thought Models." September 12, 2024. https://simonwillison.net/2024/Sep/12/openai-o1/.

Wong, Lionel, Gabriel Grand, Alexander K. Lew, Noah D. Goodman, Vikash K. Mansinghka, Jacob Andreas, and Joshua B. Tenenbaum. "From Word Models to World Models: Translating from Natural Language to the Probabilistic Language of Thought." arXiv, June 22, 2023. https://arxiv.org/abs/2306.12672.

Wright, Webb. "Coca-Cola's Pratik Thakar on the Power and Potential of Generative AI." The Drum, August 23, 2023. https://www.thedrum.com/news/2023/08/23/coca-cola-s-pratik-thakar-the-power-and-potential-generative-ai.

Wu, Tianhao, Weizhe Yuan, Olga Golovneva, Jing Xu, Yuandong Tian, Jiantao Jiao, Jason Weston, and Sainbayar Sukhbaatar. "Meta-Rewarding Language Models: Self-Improving Alignment with LLM-as-a-Meta-Judge." arXiv, July 28, 2024. https://arxiv.org/abs/2407.19594.

Xue, Fuzhao, Yao Fu, Wangchunshu Zhou, Zangwei Zheng, and Yang You. "To Repeat or Not to Repeat: Insights from Scaling LLM under Token-Crisis." arXiv, May 22, 2023. https://arxiv.org/abs/2305.13230.

Y

Yehudai, Gilad, Haim Kaplan, Asma Ghandeharioun, Mor Geva, and Amir Globerson. "When Can Transformers Count to N?" arXiv, July 21, 2024. https://arxiv.org/abs/2407.15160.

Yudkowsky, Eliezer. "MIRI Announces New 'Death with Dignity' Strategy." LessWrong, April 1, 2022. https://www.lesswrong.com/posts/j9Q8bRmwCgXRYAgcJ/miri-announces-new-death-with-dignity-strategy.

Yudkowsky, Eliezer. "My Childhood Role Model." LessWrong, May 23, 2008. https://www.lesswrong.com/posts/3Jpchgy53D2gB5qdk/my-childhood-role-model.

Yudkowsky, Eliezer. "Pausing AI Developments Isn't Enough. We Need to Shut It All Down." *Time*, March 29, 2023. https://time.com/6266923/ai-eliezer-yudkowsky-open-letter-not-enough/.

Yurman, Daniel. "Is Microsoft & OpenAI's 5GW Stargate Supercomputer Feasible?" Data Center Dynamics, April 9, 2024. https://www.datacenterdynamics.com/en/opinions/is-microsoft-openais-5gw-stargate-supercomputer-feasible/.

Z

Zhang, Edwin, Vincent Zhu, Naomi Saphra, Anat Kleiman, Benjamin L. Edelman, Milind Tambe, Sham Kakade, and Eran Malach. "Transcendence:

Generative Models Can Outperform the Experts That Train Them." arXiv, June 17, 2024. https://arxiv.org/abs/2406.11741.

Zhang, Hugh. "o1 Test-Time Compute Scaling Laws." GitHub, 2024. https://github.com/hughbzhang/o1_inference_scaling_laws.

Zhang, Hugh, Jeff Da, Dean Lee, Vaughn Robinson, Catherine Wu, Will Song, Tiffany Zhao, Pranav Raja, Dylan Slack, Qin Lyu, et al. "A Careful Examination of Large Language Model Performance on Grade School Arithmetic." arXiv, May 1, 2024. https://arxiv.org/abs/2405.00332.

Zhang, Wenqi, Yongliang Shen, Linjuan Wu, Qiuying Peng, Jun Wang, Yueting Zhuang, and Weiming Lu. "Self-Contrast: Better Reflection through Inconsistent Solving Perspectives." *Proceedings of the 62nd Annual Meeting of the Association for Computational Linguistics* 1 (August 2024): 3602–22. https://aclanthology.org/2024.acl-long.197/.

Zheng, Lianmin, Wei-Lin Chiang, Ying Sheng, Siyuan Zhuang, Zhanghao Wu, Yonghao Zhuang, Zi Lin, Zhuohan Li, Dacheng Li, Eric P. Xing, et al. "Judging LLM-as-a-Judge with MT-Bench and Chatbot Arena." arXiv, June 9, 2023. https://arxiv.org/pdf/2306.05685.

Ziegler, Daniel M., Nisan Stiennon, Jeffrey Wu, Tom B. Brown, Alec Radford, Dario Amodei, Paul Christiano, and Geoffrey Irving. "Fine-Tuning Language Models from Human Preferences." arXiv, September 18, 2019. https://arxiv.org/abs/1909.08593.

ZoomRx. "State of AI Report." April 2024. https://www.zoomrx.com/reports/FERMA_State_of_AI_Report_April_2024.pdf.

Acknowledgements

Thanks to Juan Vazquez for technical support, moral support, and graphs. Thanks to Zheng-dong Wang, Stanislav Fort, Uzay Girit, Harry Law, John Burden, Nuño Sempere, Simon Steshin, Rory Švarc, and David Mathers for helpful comments. Thanks to Rebecca Hiscott for graceful coaching and editing; it would be a lesser book without it.
Gavin Leech

Special thanks to Sholto Douglas and Leopold Aschenbrenner for teaching me much of what I know about AI, both on and off camera.

Thanks to Bryan Caplan, Tyler Cowen, and Anil Varanasi for their early, persistent, and unfailing mentorship and support for my career.

Most of all, thanks to my parents, Nita and Sanjay Patel, for sacrificing so much to give me the opportunities I have received.
Dwarkesh Patel

About the Authors

Dwarkesh Patel is the host of *Dwarkesh Podcast*, where he produces deeply researched interviews with both obscure intellectuals and the most influential figures of our time.

Gavin Leech is the cofounder of Arb Research, a consultancy that does empirical and conceptual work in various sciences. He is miscellaneous at gleech.org.

About *The Scaling Era*

How did we build large language models? How do they think, *if* they think? What will the world look like if we have billions of AIs that are as smart as humans, or even smarter?

In a series of in-depth interviews with leading AI researchers and company founders—including Anthropic CEO Dario Amodei, DeepMind cofounder Demis Hassabis, OpenAI cofounder Ilya Sutskever, MIRI cofounder Eliezer Yudkowsky, and Meta CEO Mark Zuckerberg—Dwarkesh Patel provides the first comprehensive and contemporary portrait of the technology that is transforming our world.

Drawn from his interviews on the *Dwarkesh Podcast*, these curated excerpts range from the technical details of how LLMs work to the possibility of an AI takeover or explosive economic growth. Patel's conversations cut through the noise and the hype to explore the topics compelling those at the forefront of the field: the power of scaling, the potential for misalignment, the sheer input required for AGI, and the economic and social ramifications of superintelligence.

The Scaling Era offers readers unprecedented insight into a transformative moment in the development of AI—and a revealing vision of what comes next.